Hermit Arithmetic

Hermit Arithmetic

… everything you wanted to know, but nobody would tell you!

Narrated by Saul Latyn
Transcribed by George M. King, Ph.D.

iUniverse, Inc.
New York   Lincoln   Shanghai

Hermit Arithmetic
… everything you wanted to know, but nobody would tell you!

iUniverse books may be ordered through booksellers or by contacting:

iUniverse
2021 Pine Lake Road, Suite 100
Lincoln, NE 68512
www.iuniverse.com
1-800-Authors (1-800-288-4677)

ISBN: 978-0-595-44417-5 (pbk)
ISBN: 978-0-595-88746-0 (ebk)

Printed in the United States of America

Contents

Preface

This is not a mathematics book—this is, however, a book about mathematics. Most of us think about mathematics books as textbooks with definitions, proofs, examples and sets of problems for someone to attempt to ascertain if they have or have not acquired the necessary skills to move forward. This book has very little, if any, of that structure. This book is a dialogue between two people; call them the teacher and the student. Their discussions do entail arithmetic concepts. In that respect it is a book about mathematics.

I taught mathematics for over thirty years, on every level from middle school to university math. I taught in both public and private schools. I taught overseas and in the U.S. During that time I felt the thrill and enjoyment of seeing my students learn, sometimes learning more than I taught them. I also felt the disappointment that many of my colleagues also felt when we all thought that the students should have fared better in their courses than they had. It is that almost universal disappointment that served as the genesis for this discourse.

I spent the last fourteen years of my teaching career at North Florida Community College in Madison, FL. NFCC was a wonderful place to teach, with professional freedom to experiment with teaching techniques, great interaction with other mathematics faculty, and the resources to give presentations at both the state and national level.

I began to realize (at some point in time) that difficulties and problems in learning mathematics were found across all spectra. It really made no difference if I was in a class of college-age students attempting to learn fourth grade arithmetic, in a seminar with high school teachers desiring insights into teaching the basics, or in a presentation with an audience of college professors who, like myself, were often tearing their hair out when teaching basic mathematics courses. I began to make notes on happenings in each of these events. If I was returning from a class or a seminar I would immediately record discourses that I might later feel were important for teaching the same concept another time.

Eventually these discourses became the notes that would become discussions—discussions between student and teacher. This was the origin of the Hermit as a teacher. The student, as presented here, is really a compilation of all of the groups named above. Because of the nature of this evolution, this book is far from comprehensive. There are many concepts of arithmetic that are not addressed here. The ideas that are presented here are those that people thought strongly enough about to mention them repeatedly in a mathematically based learning environment.

I began writing this as a guide for students—a guide that could give them, in much more detail than a textbook, the origins or uses of many of these concepts. This expanded to a guide for students in mathematics education and teachers of mathematics. Eventually, through my totally unscientific process of asking friends and acquaintances to read a chapter or two, I have found that this is readable and understandable for anyone from sixth grade through a college graduate degree. I have also seen that anyone in that range can also learn something in the reading.

Deepest thanks to my wife, Linda, and my son, Travis. They were by my side through all of the dark times and serious illnesses. This project would never have reached fruition without their assistance.

Prologue

Before we start with this story of the "Math Hermit" we should all know as much as there is about this person. There is not a lot to know, as his history is vague at best. But, … There are some important things that we should all be aware of before we continue.

The Hermit himself related many of these details to me, and many others were related to me by other Hermit students. Where they originated is really irrelevant, since we are not sure of the truth or falsehood of any of them. The main reason for beginning the account of the Hermit this particular way is that we will all have a common starting point, something that will become increasingly important as we continue to amass information.

Early History

The earliest known details of the Hermit history begin in the Florida Panhandle, or at least in Northern Florida. While none of us I really sure of his real name, with his assistance we are able to ascertain many details of his early life.

This early history of his past entails his attendance of college and graduation in those early years. He attained a bachelor's degree in mathematics from a northern Florida University at this time. After doing so he began to teach mathematics in a variety of high schools. He did this for a couple of years, and then began to travel west across the Gulf States. He traveled west through Alabama, Mississippi, and finally to Louisiana. He enjoyed the ambience of Louisiana and decided to stay there.

Louisiana Days

His real discovery in traveling west was not in Louisiana itself as much as it was in his discovery of New Orleans. It was as if New Orleans and the Hermit were made for each other.

While in New Orleans he developed his basic techniques for one-on-one tutoring of mathematics. His method for doing this was to spend each afternoon at the Riverside Park just off of Front Street, tutoring a variety of students at a variety of levels in mathematics. These students were able to find him because he was concurrently enrolled in Master's Level mathematics courses at one of the university's in New Orleans. He spent the morning in classes, the afternoon in the riverside park, and the evening enjoying the combination of Cajun jazz and it's mathematical basis.

In addition to all of the above the Hermit also decided to participate in a college varsity sport. He chose football, for which he was ideally suited. He was tall and gangly as opposed to most college football players who were shorter than he and also heavier. He ran like a gazelle among rhinos. His long loping strides were difficult to stop, and he was always giving opponents a target leg and then recalling it. He was not good enough to play on every down, but was good enough to play on kick-off teams, teams on which he was difficult to block or tackle. It is said that he was the only person who could consistently tackle Forrest Gump on kick-offs. It is not that he was better than Gump but he could not be blocked and stopped before he tackled Gump.

He earned his Master's Degree in mathematics while in New Orleans. He decided that it was time, once again, get on the road and explore his small part of the world. This time he traveled north and landed in the vicinity of Little Rock.

Little Rock and West....

Little Rock proved to be an excellent place to hone his teaching skills. There were teaching jobs available and many opportunities to practice his one-on-one tutoring skills. He eventually decided that he no longer wanted to teach in a large class environment, and wanted to strictly tutor. He had made enough contacts in Little Rock to be able to do this and still make enough money to live on. He was able to do this for a couple of years before becoming bored with it. He was not bored with the tutoring, but bored with what the environment had to offer. He liquefied all of his assets (little as they were) and began moving west.

Eventually his travels led him to New Mexico. There he found a wonderful and primitive existence on the edge of the desert. He was far enough from civilization to be comfortable and close enough to encourage students to come to him for tutoring in the basics of mathematics. Perhaps the strangest thing about his life story is that he was able to find, and purchase, some cliff caves that had been constructed by Native Americans hundreds (maybe thousands) of years previous. He purchased a small cluster of these caves that provided him water, cool air in the summer, warmth in the winter, and an abundance of other natural amenities. His needs were little, and he found that he could trade his home-grown vegetables and batteries at the local general store (only 10 miles away) for any goods he needed.

It was not long before he developed a 'cult' following of students in need of (or in want of) tutoring. They would hike to the Hermit Cave, and once there would discuss with him whichever mathematics topic they desired. There were never any time limits, rooms, etc. I became, first only out of curiosity, one of his students. I developed a relationship with him unlike any other I have ever had with another teacher. This small volume is a compilation of many of our discussions.

Saul Latyn

Session 1—Hermit Addition

What exactly is addition?

One day I happened to visit the Hermit and as we were talking I brought up the topic of addition. He asked me to define, in my own words, what "addition" was.

"Addition is, I think, finding the sum or total of two numbers." I stated.

"So, you think that you need numbers in order to add? You do not think that you can add without using numbers?", was his reply.

"Well, yes, I guess I do need numbers."

"Well, no, I guess that *I do not need numbers.*"

"Please explain how you can possibly add without using numbers."

He was, as always, patient and methodical in his explanation. He got up from his stool and walked across the cave to the large trunk in the corner of the room. He opened the trunk and rustled around inside, searching for something. He finally found what he was looking for. He came back with a purple silk bag trimmed in yellow, one that had in a previous life held a bottle of some sort. He returned, sat on his stool, held the bag in his lap, and simply stated, "Now, I'm going to add, and I am not going to use any numbers."

He reached into the bag and withdrew some smooth, shiny pebbles. He proceeded to carefully put these into a pile on the ground before him. He then said, "Now I am going to add, or perform the operation of addition." He put his hand back into the bag, removed some pebbles, and put these also onto the pile. "There, I just *added* to the pile."

"That's not fair," I said. "You didn't add. You don't know how many are there. You didn't count the pebbles. You don't know the answer."

"Was there not a pile on the ground? Did I not 'add' to that pile?

"Well … yes … but…."

"No 'buts' here. I performed an addition."

"OK, so what *is* addition?"

"No, *you tell me.* You admit that I added. You saw me do it. Maybe it is easier to define addition if you tell me what it was that I did, or what I had."

"You had a pile of pebbles, and you placed more pebbles on the pile."

"That's correct. You need to start with *stuff,* and you place more of that *stuff* on the pile, and you have 'added' to the *stuff.*"

"But … you don't know the answer!", I stated boldly.

The ANSWER

"You are correct that I don't know the answer. But … I did *perform an addition*. And … I did not have to use numbers in order to do the addition. Your problem is that you think in terms of 'finding an answer' rather than the definition of the process."

"Aren't most people interested in 'finding the answer'?"

"Yes, they are, and in that case we need to determine how many were in the initial pile, and how many were 'added' to the pile."

"Then, that is addition."

"NO, that is the result of the addition."

"Well, OK, there's no 'trick' to that, is there?"

"No, no tricks here. Mathematics does not involve tricks. Mathematics involves analysis of the problem with a logical solution. Let's look at an addition problem and determine, without seeing the 'piles', what the answer you want, is."

The COMPUTATION

"Let's add two numbers. Let's suppose, for example, that the initial pile contained 234 pebbles, and to that amount you 'added' 567 pebbles. You would now want to know how many pebbles were in the pile", stated the Hermit. He proceeded to pick up a stick, which happened to be right next to him, and drew the following in the sand.

$$234$$
$$+567$$

"I can do this", I said confidently.

I began, "Seven and four is 11—write 1, carry 1. 6 and 3 is 9 plus 1 is 10—write 0 carry 1. 2 and 5 is 7 and 1 is 8. So …

$$11$$
$$234$$
$$+567$$
$$801$$

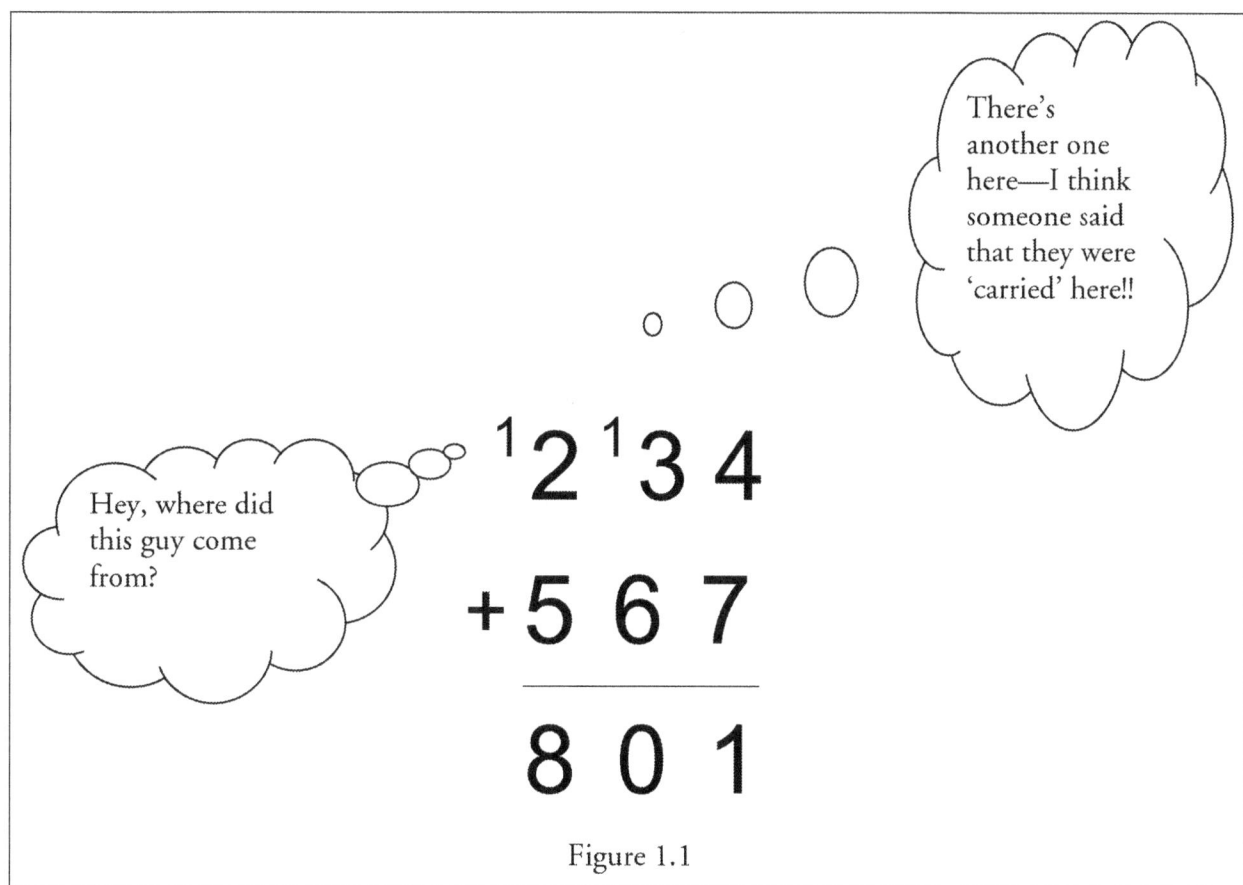

Figure 1.1

"What is this 'carry' stuff that you are doing, and why do you even do it?" he inquired.

He continued, "Why do you put those little '1s' up in the air. I don't understand that!"

"I'm carrying!" I stated, quite exasperated. "You always have to carry when you have more than ten."

"You have this *need* to carry. Where do you carry them? Do you put them in a sack—in a box—in a plastic bag—just how do you 'carry' numbers?

"You know what I mean. You carry them to the next column. That's just the way you do it!"

"That's not the way that I do it."

"Then you have to show me how you do it."

"Thanks, it's time to learn another way to add! There's more than one way to *skin this cat* called addition. Here's the question:

$$234$$
$$\underline{+567}$$

The SOLUTION

"You recall, of course, our previous conversation regarding place values and expanded numbers, do you not?"

"Yes, of course I do!" I replied with a certain amount of hurt in my voice.

"No need to get upset, I was just asking," he replied. He then continued, "So, I can re-write these numbers using expanded notation. Let me do that here."

$$234 = 200 + 30 + 4$$
$$+567 = 500 + 60 + 7$$

"Now," he continued, "I can add these."

$$234 = 200 + 30 + 4$$
$$+567 = 500 + 60 + 7$$
$$700 + 90 + 11$$

"Now I can easily see the answer. I add 700 and 90; thus, the answer is 790. To this I add 11, and the final answer is 801!"

"But," I interrupted, "shouldn't you have carried since you had more than 9 in the unit column?"

"If I were adding your way I would have. But, adding eleven is simple enough to see—I almost have that many fingers! Let's look at it again.

"Now, when you add 7 and 4 the answer is 11. when you add 6 and 3 you are not really adding 6 and 3, but what are you adding?"

"You are adding 60 and 30!!"

"Correct! And 2 and 5 is ..."

"200 and 500!"

"Yes. So the solution is the following:"

$$
\begin{array}{r}
234 \\
567 \\
\hline
11 \\
90 \\
700 \\
\hline
801
\end{array}
$$

"Let's look at another example," he said as he smoothed the sand. He proceeded to draw another problem on the floor.

```
   1245
   3456
   ____
     11
     90
    600
   4000
   ____
   4701
```

"You may notice that this method is a little longer, but I am using my concepts of *place value* and our *decimal system of numeration* we earlier said was so important to learning mathematics. Another positive aspect of this method is that I can simply 'read' my answer: I can easily add 4000, 600, and 90 to obtain 4690. To this I simply add 11 for the final result, 4701!! I can even add from left to right, just as I read."

"No way," I said, stunned!

"Watch, you are going to *like* this."

He proceeded to put down another problem, and explain as he wrote the answer.

```
   5234
   4568
   ____
   9000
    700
     90
     12
```

"5000 and 4000 is 9000; 200 and 500 is 700; 30 and 60 is 90; and, finally, 4 and 8 is 12. Now I can simply 'read' the answer: I have 9790, to which I add 12, and the answer is 9802!"

```
   5234
   4568
   ____
   9000
    700
     90
     12
   ____
   9802
```

"Let me give you one more to do by yourself." He was smirking now that he knew that I was hooked by this method. The next addition almost flew out of his head.

$$4574$$
$$+8556$$

I spoke as I wrote. "4000 plus 8000 is 12000. 500 and 500 is 1000. 70 and 50 is 120. 4 and 6 is 10."

$$4574$$
$$+8556$$
$$12000$$
$$1000$$
$$120$$
$$10$$

"The answer is 13,130!

So ... carrying is bad?

I needed to know what he *really* thought about carrying, and I just had to ask.

"So, carrying is really a bad idea."

"No," he replied. "Using your method of carrying is used by a large majority of people for additions. It is compact and the process is condensed to one line below the problem, not spread down the whole page. There are many times that I use your method of *carrying* when I do additions."

"But...."

"But ... you should know why you are doing when you are using mathematics and its operations. I have tried to show you a method that shows you *why you do what you do*. If you always remember that there are reasons you do the things you do in mathematics, then the process becomes one that is understandable. Mathematics should not, in anyone's mind, simply be a lot of tricks and moving numbers around without knowing any reasons for what you do. The more you understand these reasons behind the mathematics you do the more the 'mystery' will be removed from mathematics."

He stopped there and simply sat staring at me. It seemed as if he was encouraging me to say something, but I could think of nothing more than trying to absorb all that had taken place that day. Finally I said,

"I can come back another time."

"That would be good. You know where to find me."

"I'll bring some batteries."

"Make sure that they are fresh."

Session 2—Hermit Subtraction

The concept of Subtraction

"So …," I blurted out. (I always hated when I did this, but now was the time. We had been talking about the geology of the valley and caves for what seemed an eternity. I didn't think that I needed to know more about the rocks around me.) "What about subtraction?"

The Hermit snorted and groused. He was squatting on his haunches at the entrance to the cave, pointing out the various geological features of the valley below and the caves in the walls of the valley. Squatting and low-riding Paco shorts did nothing for his appearance. Finally he stood, tugged on his shorts, and sat on the stool that he had previously abandoned.

"The concept of the operation of subtraction in arithmetic (in which only whole numbers are used) can be very simple or very confusing. The whole idea of subtraction is to start out with an amount of something, and then take away some of these things. The result of the subtraction is what remains from the original amount. You do realize this?"

"Yes, of course."

"Good. Then, here is what we have."

He proceeded to pick up a stick and draw in the sand between us.

"So," he continued, "If we start with an amount, say 43, and we remove 12 from the 'pile of things', we will be left with 31 in our 'pile of things'. Is this correct?"

"I would agree."

Start with a certain amount.	Remove this many.	What is left of the original

"Mathematically we would see this." He quickly scribbled in the sand.

$$43$$
$$\underline{-12}$$
$$31$$

"I agree again." I had a feeling that this was *too easy*. He was leading me into a *Hermit trap*. But … I knew that it would be a *good* trap. It would be a trap that would make me think, make me learn.

He continued, smoothing sand and drawing as he spoke. "In order to visualize this process we can really look at these amounts:" He again smoothed some sand and drew the following:

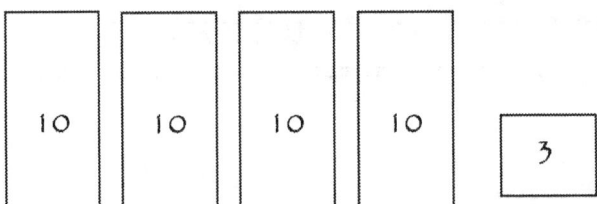

"This is our representation of forty-three, consisting of four groups of ten and one group of three. And here … here is our representation of twelve."

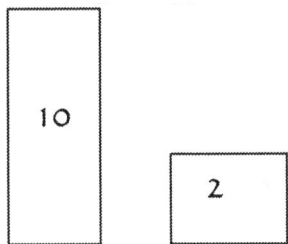

"This has one group of ten and one group of two. If we remove one of the 10s and two from the group of three, we will be left with three groups of 10 and one unit. This is what was really written in the original problem. You can see that in the problem statement as originally written:

$$
\begin{array}{r}
43 \\
-\,12 \\
\hline
31
\end{array}
$$

Subtraction with "borrowing"

"I agree with that. It all makes perfect sense." I said. I knew that the trap was being set. I smiled to myself as he continued.

"If we look at another example we can see that it is not always that easy. Suppose we have 45 objects from which we wish to subtract 19. We would have the following:"

Smoothing the sand and drawing, almost in one motion, he drew.

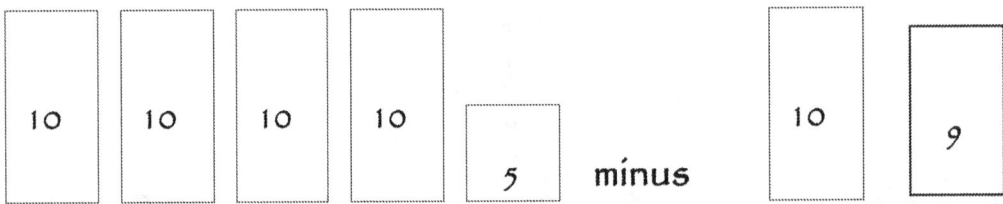

Again, he continued, drawing as he spoke. "Since we cannot remove 9 units from the 5 units, we must re-group our subtraction problem as follows:"

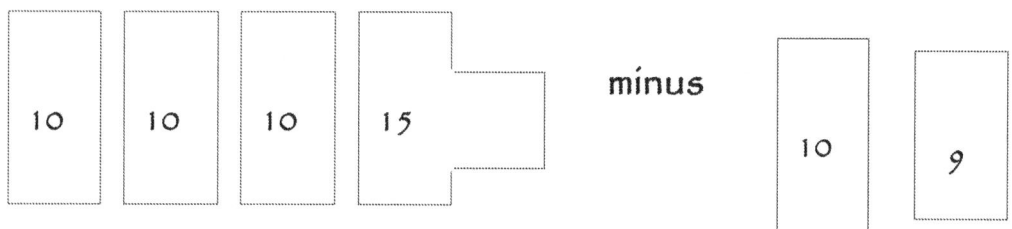

"We can now take one 10 from the three 10s and 9 units away from the group of 15 units. Thus, we are left with 26 from the original problem! Using symbols we 'borrow' one of the tens and place these 10 units into the group with the original 5 units. We can show this in the following way:"

$$\begin{array}{r} {}^3\mathbf{4}\ {}^1\mathbf{5} \\ -\ \mathbf{1}\ \ \mathbf{9} \\ \hline \mathbf{2}\ \ \mathbf{6} \end{array}$$

"But …" he continued, "this really becomes tedious when we have to deal with larger numbers, such as in the following problem."

$$\begin{array}{r} \mathbf{2\,0\,1\,0\,4} \\ \mathbf{-2\,3\,4\,5} \\ \hline \end{array}$$

"The solution looks like this:"

$$\begin{array}{r} {}^1\mathbf{2}\ {}^{9}_{10}\mathbf{0}\ {}^{10}\mathbf{1}\ {}^{9}_{10}\mathbf{0}\ {}^1\mathbf{4} \\ -\quad \mathbf{2}\ \ \mathbf{3}\ \ \mathbf{4}\ \ \mathbf{5} \\ \hline \mathbf{1}\ \ \mathbf{7}\ \ \mathbf{7}\ \ \mathbf{5}\ \ \mathbf{9} \end{array}$$

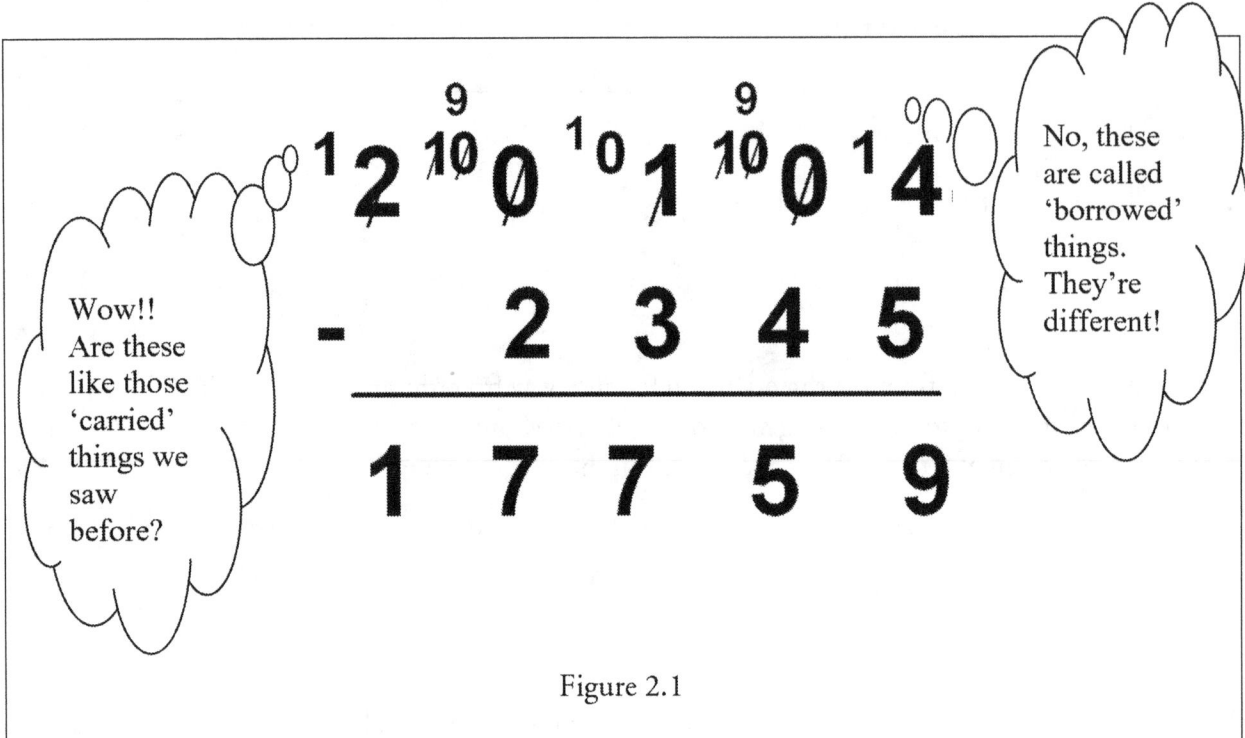

Figure 2.1

"Yes," I interrupted, "That's the way you have to subtract. You have to *borrow* when the top number is smaller than the bottom number. But … everyone knows this!"

"You may know this, and so may most others. But … what I have written is more than sufficient to make anybody confused. There is a much simpler way to subtract, and, *even better*, you do not have to borrow … or subtract!"

Subtraction without subtracting

"Com'on," I said, "You can't subtract without borrowing *sometimes*. And … you can't subtract without subtracting. That just doesn't make sense!"

He continued, nonplussed, "Let's examine our original subtraction problem. Recall that it stated the following."

Smooth and draw, smooth and draw.

$$43 - 12 = 31$$

"We started with a pile (or box, or bag) of 43 objects. From this group we removed 12. The result was that 31 objects remained from the original group. That was our answer. How do we *know* that we are correct? The answer is simple: return the 12 we removed and see if there are again 43 … **or add the 12 back to the remaining 31 and verify that we have 43**. This is the most common way to check your subtraction problems:"

"Yes, that is the way that you assure that you are correct."

He scribbled the following.

$$12$$
$$+\,\underline{31}$$
$$43$$

"We are again going to look at this problem," he stated, "but from a different point of view. Here is the original problem:"

$$43$$
$$-\underline{12}$$

"Recall our checking of the problem. We can now ask, what must I add to 2 in order to have 3, and the answer is 1. What must I add to 1 to have 4, and the answer is 3. So...." He scribbled again in the sand.

$$43$$
$$-\underline{12}$$
$$31$$

He flipped down his stick with a wave of his hand. He smiled smugly as he stated, "*Any subtraction problem can be solved by addition!*"

I gotta try this," I said. "Let me try a subtraction without subtracting."

"OK. Here is one for you to try."

He cleared the sand and wrote the following for me to try.

$$567$$
$$-\underline{135}$$

I muttered to myself as I solved the problem: "First, ask what must I add to 5 to obtain 7; next, what must I add to 3 to obtain 6; last, what must I add to 1 to obtain 5? The answers to these questions are 2, 3, and 4, respectively. Thus,"

$$567$$
$$-\underline{135}$$
$$432$$

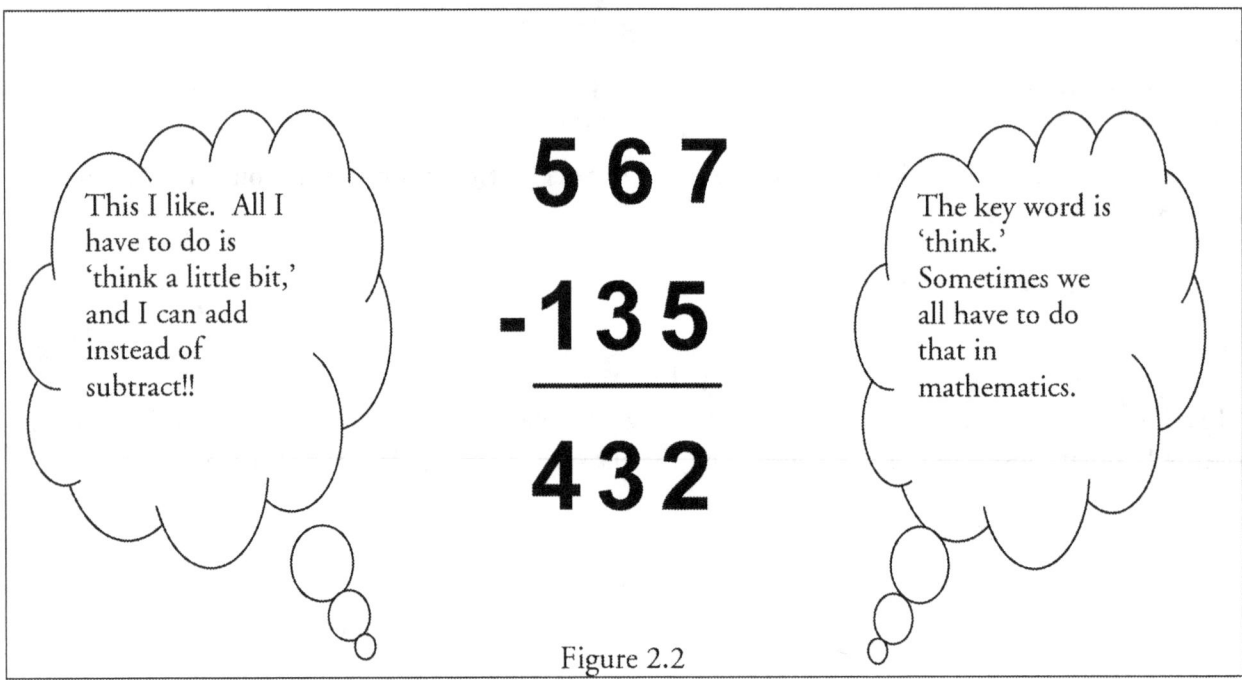

Figure 2.2

"Now comes the fun part," he said with a smile. "You never have to borrow! We'll look at one of our examples to demonstrate this. We first looked at the problem 45 − 19." Smooth and write.

$$\begin{array}{r} 45 \\ -\ 19 \\ \hline \end{array}$$

"There is nothing we can add to 9 to get 5, but … we can add 9 and 6 and the result is 15. We will do this, and carry '1' to the next column."

$$\begin{array}{r} 45 \\ -_119 \\ \hline 6 \end{array}$$

Since we carried the '1' into the 10s column, we now have a 2 under the 4, and then we can add another 2 to the one already written to get an answer of 4. Hence,

$$\begin{array}{r} 45 \\ -_119 \\ \hline 26 \end{array}$$

"I really like this. I can subtract without borrowing … or subtracting." I could no longer hold back my enthusiasm.

"We looked at an example earlier in which there was a lot of borrowing." He stated this bluntly. He then proceeded to draw in the sand the exact problem we had done earlier. I still do not know how he keeps all of these problems in his head. But … he does.

"Here is the problem we did earlier."

$$20104$$
$$-\ 2345$$

Recall that the solution looked like this:

$$^1\mathbf{2}\,^9_{10}\mathbf{0}\,^0_{10}\mathbf{1}\,^9_{10}\mathbf{0}\,^1\mathbf{4}$$
$$-\qquad 2\ \ 3\ \ 4\ \ 5$$
$$\overline{\qquad 1\ \ 7\ \ 7\ \ 5\ \ 9}$$

"Now, let's look at this same problem and do it without any subtraction and borrowing. First, notice that 5 + 9 = 14. Thus, we write the 9 in the units column and carry the 1."

$$201\ 04$$
$$-23_{\,1}45$$
$$\overline{\qquad\qquad 9}$$

"There is now a 5 under a zero in the 10's column, and 5 + 5 = 10. Write 5 and carry 1."

$$20\ 1\ 04$$
$$-2_{\,1}3_{\,1}45$$
$$\overline{\qquad\qquad 59}$$

"There is now a 4 in the 100s column under a 1, and 4 + 7 = 11. Write 7 and carry the 1.

$$20\ \mathbf{1}\ 04$$
$$-\ _{\,1}2_{\,1}3_{\,1}45$$
$$\overline{\qquad\qquad 759}$$

"Finally, there is a 3 under a 20, and 3 + 17 = 20. We are done here!"

$$20\ \mathbf{1}\ 04$$
$$-_{\,1}2_{\,1}3_{\,1}45$$
$$\overline{\qquad 17\ 759}$$

Or ... do it this way!

I couldn't believe it. We had just subtracted two numbers ... and ... we didn't subtract ... and we didn't borrow.

"This is great," I said. "It can't get any better, can it?" I said this with some hope in my voice that it really could get better.

"You want more?" He questioned. "Well ... let's examine this process just one more time. Maybe you will prefer this method. We begin by looking at a subtraction problem in general. It looks like this:" Smooth and write ... smooth and write. I don't think anyone could smooth sand like the Hermit.

$$
\begin{array}{r}
M \\
- \underline{S} \\
D
\end{array}
$$

"We begin with a certain quantity, called *M*, subtract from it a quantity, called *S*, and the result is the difference, called *D*. In order to check the answer, we add *S*, the amount taken away, to *D*, the answer. The check would look like this:"

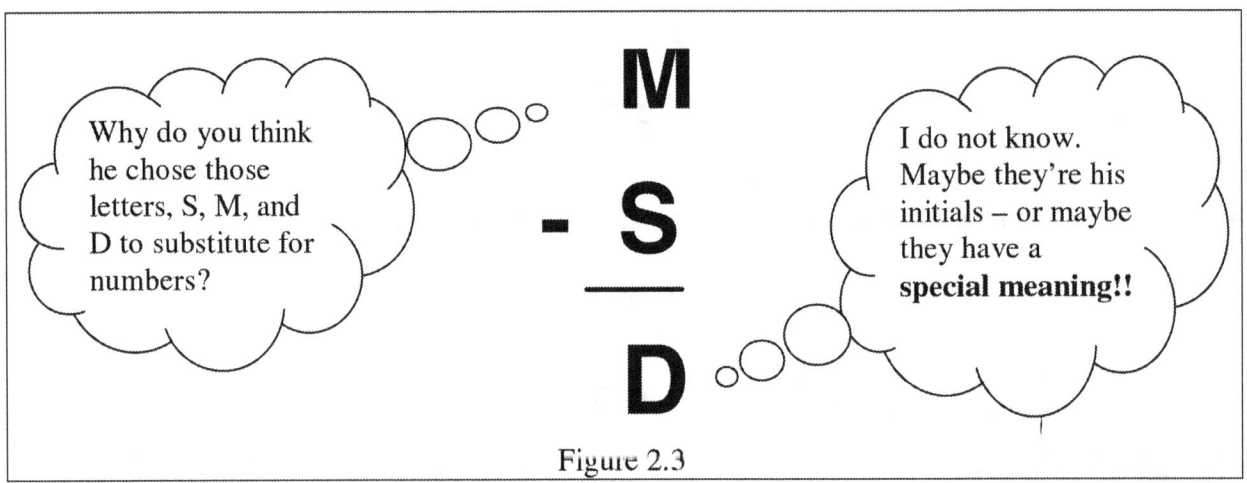

Figure 2.3

$$
\begin{array}{r}
S \\
+ \underline{D} \\
M
\end{array}
$$

"We will begin by looking at a simple example: 67-32. Comparing this to the two problems above, we are looking at the following:

$$
\begin{array}{r}
67 \\
- \underline{32} \\
D
\end{array}
\quad \text{or} \quad
\begin{array}{r}
32 \\
+ \underline{D} \\
67
\end{array}
$$

I was not about to interrupt the Hermit. I knew from past experience that he was on a roll.

"Now, focusing on the problem on the right, if we add 5 to 2 the result is 7, and if we add 3 to 3 the result is 6! Hence, 32 + 35 = 67, and 67 − 32 must be 35."

```
    67          32
  - 32   or   + 3 5
  ─────       ─────
    D           67
```

"Wow," I exclaimed. "Can I try one … one that entails borrowing?"

"Definitely. I was hoping you would ask. Here is a problem which needs borrowing. Let us see if you can do it without borrowing … or subtracting."

"Whoa …," I thought. "He's going to give me a subtraction problem that involves borrowing. And … I am supposed to do it without borrowing … or subtracting!!" This should be a challenge.

"Here is your problem, one that would ordinarily entail *borrowing*." He scrawled the problem in a clear area of sand.

```
    473
  - 175
  ─────
```

I looked at the problem for a while. I wanted to do it correctly on the first try. I muttered while I wrote. "First, I will re-write the problem as an addition problem, only I will be *missing an addend.*

```
    175
  + ───
  ─────
    473
```

"I can see that 5 + 8 = 13. I can write down the 8 and carry the 1." I wrote this down carefully, knowing that I was under his close scrutiny.

```
   1'75
  +   8
  ─────
   473
```

"Now," I proceeded, "moving left to the 10s column, there is now an 8 on top, so 8 + 9 = 17. We will write the 9 and carry the 1."

```
  ¹1'75
  +  98
  ─────
   4 73
```

"Now," I said softly, thinking aloud, "in the 100s column there is a 2 on top, and 2 + 2 = 4. Thus, we have the following:"

$$\begin{array}{r} {}^1{}1{}^1{}7\,5 \\ +\ 2\,98 \\ \hline 4\,73 \end{array}$$ ⟵— Answer!

"So," I stated proudly, "I have just shown that 473 − 175 = 298! And … I did it without subtracting *or* borrowing!"

"You did an excellent job. It is *so* easy when you don't have to subtract and borrow. I honestly don't know the last time that I did either of those!"

Hermit Note:

Here is another example, simply because I think that another would be helpful:

$$4651 - 3380$$

We will solve it by changing it into a related addition problem.

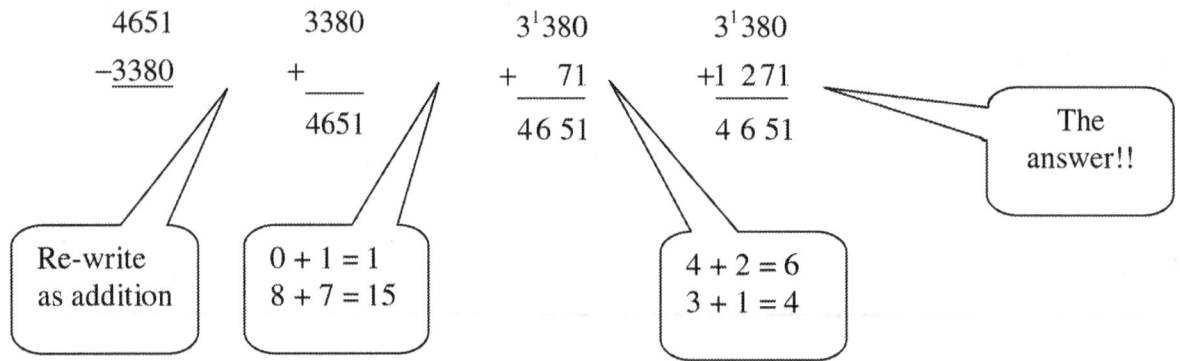

Here is an example that would entail a lot of *borrowing*. We will use the addition method to solve the subtraction problem.

We have just verified that $800006 - 345678 = 454328$.

I, the Hermit, hope you enjoyed our trip through subtraction. There is nothing wrong with *borrowing*. But ... it is much easier to translate a subtraction problem into a related addition problem (at least for me).

Session 3—Hermit Multiplication

My Question—and the Hermit Answer

"Why, do you think, that all multiplication tables go only up to twelve?" I said one day when visiting my friend the Hermit.

"I'm not sure." He began. "Maybe it is because whoever began writing these tables didn't feel that we need to know more than our multiplication facts over twelve. Or perhaps, the people who write these tables have triskaidekaphobia!"

"Triska..what?"

"Triskaidekaphobia, the fear of the number thirteen!" He then continued, "Maybe the first tables were written by the Babylonians."

"The Babylonians?" I inquired.

"Yes, he continued. "They used a base 60 for their mathematics. We have developed a system that uses a base 10, called a decimal system. We do, however, still have leftovers from a base 60 system."

"Such as what? We use base 10 for all of our mathematics."

"Such as … have you looked at a clock lately. We do not have 10 or 20 hours, but have two half days, each of which has 12 hours. And, each hour has 60 minutes, and each minute contains 60 seconds. I'm sure this is a holdover from the Babylonians.…

"I think that I should invent a clock with a 20 hour day, and each hour will contain 100 minutes, and each minute will contain 100 seconds. Wouldn't that be convenient?"

"Well, would it still be a real day?" I asked quizzically.

He continued as if never interrupted, "The size of the hours, minutes, and seconds would have to change. Since there would only be 20 hours, they would have to be longer than the hour we are used to. I will have to give this some more thought.

"Back to your original question of why multiplication tables go up to twelve. That is not really necessary. If you think about multiplication, and performing a multiplication, you only need to know single digit multiplications! You do recall the definition of multiplication?"

The Meaning of Multiplication

"Of course I do," I stated boldly. "Multiplication is simply a short-cut for performing a repeated addition. For instance, five times three is simply the same as adding five threes together. Look."

I smoothed some sand in front of me and wrote the following with a small stick lying at my feet.

$$5 \times 3 = 3 + 3 + 3 + 3 + 3$$

"That's correct," he stated. "But, we do not always want to write multiplications as addition problems. If I had the problem 32 times 19 I would not want to write thirty-two nineteens and then add them! That's where multiplication comes in handy. It's a good thing we have this little short-cut!"

Do The Math …

"Can I show you how I do that problem?" I asked. "I mean the 32 times 19."
"If you wish. I will just watch." He leaned back and settled in on his stool.
I wrote the following in the sand.

$$
\begin{array}{r}
32 \\
\times 19 \\
\hline
\end{array}
$$

I began to do the multiplication, explaining as I wrote: "Nine times two is eighteen, write the eight and carry the one. Nine times three is twenty-seven: add the one and write twenty eight."

$$
\begin{array}{r}
\overset{1}{3}2 \\
\times 19 \\
\hline
288
\end{array}
$$

"Now," I continued, "one times two is two and one times three is three, so…."

$$
\begin{array}{r}
\overset{1}{3}2 \\
\times 19 \\
\hline
288 \\
32 \\
\hline
\end{array}
$$

"And now I simply add to determine the solution. Eight and two is ten, carry the one, and three and two and one is six. Here it is."

$$
\begin{array}{r}
\overset{1}{3}2 \\
\times 19 \\
\hline
{}^{1}288 \\
32 \\
\hline
608
\end{array}
$$

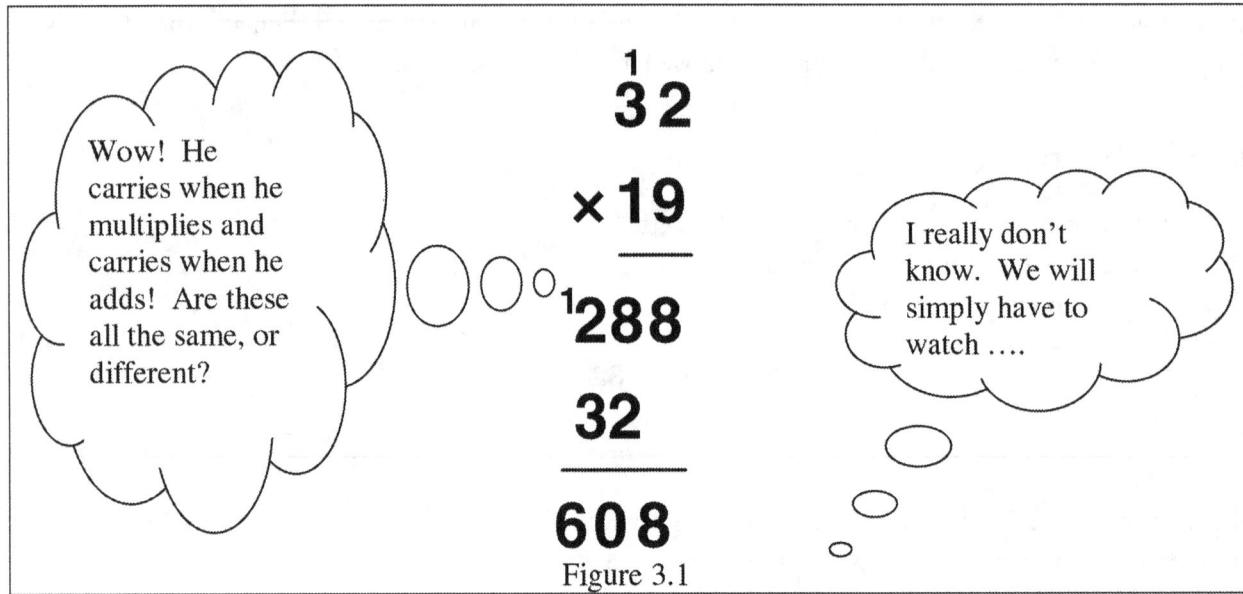

Figure 3.1

"Very nice." A compliment from the Hermit! He then continued, "Recall what I said earlier, that we really don't need multiplication tables that go above *nines*. When you do a multiplication problem the way you did it you always only do single digit multiplication. Very nice, very simple. I like it."

"Well, thank you. It is really rather simple."

"I do, however, have some questions about your problem and the solution. You don't mind my asking, just to make it more clear for me exactly what you did?"

"Not at all," I said confidently.

He began firing questions at me faster than someone spitting seeds when eating a watermelon. "Why did you carry the one over the three? When you multiplied one by two, why did you put the two under the *middle* eight? Why is there nothing under the right-most eight?"

"Wait, wait," I begged. "Give me a chance. The answer to all of your questions is because that's the way it is done. That is the way to arrive at the right answer."

"But," he continued, "There should always be a reason for doing things in mathematics. The correct answer is nice to arrive at, but you should also know why you are doing the mathematics correctly. You can be sure that if you do something when performing mathematical operations that there is a good, and logical, reason for doing these things."

I began to see that I was just doing things without really knowing why. I could see where this was going when I led him to the next step in this discussion. "OK, I can't really explain this process. I just know that it is right. Would you like to show me *why we do what we do* when multiplying?"

... Be The Math

"It would be my pleasure. We will have to go a ways back here to get to the real meaning of your multiplication. You recall that we said in our decimal system of numbers we can write numbers with expanded notation. I can do that here with your two numbers. Watch." He wrote this on the floor of the cave.

$$32 = 30 + 2$$
$$19 = 10 + 9$$

"Now, the three really represents thirty and the one really represents ten. With this in mind I will do the same problem, only attack it a little differently."

I watched as he smoothed what he had written, found a new stick, and proceeded to write the problem again on the floor.

$$\begin{array}{r} 3\,2 \\ \times\,1\,9 \\ \hline \end{array}$$

"Now," he stated as he wrote, "nine times two is eighteen.

$$\begin{array}{r} 3\,2 \\ \times\,1\,9 \\ \hline 1\,8 \end{array}$$

"... and nine times thirty (which is really what the three represents) is 270.

$$\begin{array}{r} 3\,2 \\ \times\,1\,9 \\ \hline 1\,8 \\ 2\,7\,0 \end{array}$$

"Notice that when you multiplied nine times thirty-two you got an answer of 288, which is the same as I have."

He was right; I was looking at my solution as he said it.

$$\begin{array}{r} \overset{1}{3}2 \\ \times\,19 \\ \hline {}^{1}288 \\ 32 \\ \hline 608 \end{array}$$

"Now I will finish my problem. I have to multiply one (which is really 10) by two. The answer is twenty.

$$
\begin{array}{r}
3\,2 \\
\times\,1\,9 \\
\hline
1\,8 \\
2\,7\,0 \\
2\ 0
\end{array}
$$

"Now I multiply one (which is really ten) by three (which is really thirty). My answer is 300.

$$
\begin{array}{r}
3\,2 \\
\times\,1\,9 \\
\hline
1\,8 \\
2\,7\,0 \\
2\ 0 \\
\hline
30\ 0 \\
\hline
\end{array}
$$

"Notice that when I now add my answer is the same as yours. Do you now see why when you placed the two under the eight you were doing it correctly?"

"Yes, of course," I said, pleased. "It was really a twenty, not simply a two!"

"Precisely."

"But … your way can get awfully long sometimes. Is my way so bad?"

"No, once you know why you are doing it that way. You should always ask yourself *why* you are doing what you are doing. If you can explain it to yourself, then you are beginning to understand the mathematics."

I was hooked now. "Can I try one your way?"

"Sure. Pick your own problem and see if you can do it correctly."

I thought for a minute. I wanted one that would challenge me, but I also wanted to do it correctly. Finally, I said, "OK, here I go." I wrote the following.

$$
\begin{array}{r}
3\,4\,5 \\
\times\ \ 2\,6 \\
\hline
\end{array}
$$

"Now," I began, writing as I was thinking aloud, "six time five is thirty, six times forty is 240, and six times three hundred is 1800.

$$345$$
$$\times\ \ 26$$
$$\overline{30}$$
$$240$$
$$1800$$

"And, twenty times five is 100, twenty times forty is 800, and twenty times three hundred is 6000.

$$345$$
$$\times\ \ 26$$
$$\overline{30}$$
$$240$$
$$1800$$
$$100$$
$$800$$
$$\underline{6000}$$

"Now I simply add to find the final answer.

$$345$$
$$\times\ \ 26$$
$$\overline{30}$$
$$240$$
^{11800}
$$100$$
$$800$$
$$\underline{6000}$$
$$8970$$

"And my answer is 8970. Well...."

"Nicely done. And now you know why the numbers are what and where they are. There are no *half written numbers* here, nor are there any *mysterious ways of putting numbers in certain columns* without knowing why. It is all here, and it is clear. If you, indeed, had made any mistakes in multiplication they should have been easy to find. Compare this to your other method."

He quickly re-did the problem using the technique I had shown him. His final solution looked like this.

$$
\begin{array}{r}
\overset{2}{3}\overset{\overset{1}{3}}{4}5 \\
\times\ \ 26 \\
\hline
2070 \\
690 \\
\hline
8970
\end{array}
$$

"Notice," he said, "that I arrived at the same answer. But … it is much more difficult to determine exactly what I did in each step. Had I made an error it would be much more difficult to detect."

"So," I began, "how do I know when to use the short method and when to use the longer, but easier to see, method."

He thought for a few seconds before beginning. "You now know the longer method, and you know why you do what you do in using the shorter method. It is up to you which method you use. One is longer, and may take more time. One is shorter but more difficult to check. Only you can determine which method to use. You have to have the confidence to know what you are doing, and that you are doing it correctly. Nobody can make these choices for you.

"All this multiplication is making me hungry. Maybe I'll make a casserole for dinner tonight."

Session 4—Hermit Division

An Introduction to Division

I had not visited with the Hermit for some time, due to a variety of reasons. Sometimes you have to do what you have to do. I did not even know if he would remember me when I arrived at the entrance to the cave.

"Well," he said, "it has been a long time, my friend."

"Yes," I replied, "but there are other things that I *had* to do, and was unable to come here."

"Quite all right. I have been busy. This season has kept me busy. This season *always* keeps me busy. I have been putting up some canned fruits and vegetables—preparing for the future, I call it.

"If I correctly recall, the last time you were here you were questioning the operation of division, and what it means and how it is performed correctly."

How he could remember that I have no idea, but ... he did! I had a difficult time remembering what I had for breakfast yesterday.

"Talking about some things, including division, on an empty stomach is not an easy task," he stated. "I will get us some refreshment before we begin."

He moved into the depths of the cave, leaving me alone at the entrance. I found a stool and a nice, long stick that would be ideal for writing in the sand. I simply sat back and enjoyed the warmth of the rising sun while waiting for his return.

The Hermit seemed to simply 'appear' back at the cave entrance. I had not even heard him approach! He had a small tray with two glasses filled to the top and a plate with some cookies.

"There is nothing like fresh papaya juice to wake-up one's gray matter early in the morning. I have tried a variety of different juices, but this seems to stimulate me to thinking like no other. I hope you enjoy it. I also brought some fresh oatmeal cookies. Yesterday I had oatmeal for breakfast, and thought that this would be a nice and 'different' way of having a morning oatmeal."

"Thank you." I said. We then sat quietly for a while as we ate the cookies and drank our juice. We simply watched the sun rising and shortening the shadows that crept across the valley floor below.

Finally ... "So," he began, "division is our topic of the day. You have, I assume, what you consider to be a reasonable definition of division?"

"Well ... yes. I guess. Simply put, division is seeing how many time one number goes into another number. You see, four goes into twenty-eight seven times, so twenty-eight divided by four is 7."

I then proceeded to pick up my stick and write in the sand: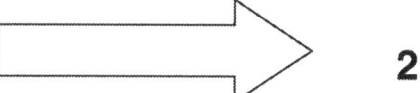

"I still have a problem with your dependence on numbers." He stated. "You want to deal with these operations, with mathematics in general, as though it is some abstraction that has no connection to the real world.

I have this need to 'feel', to 'touch,' the problem. You said that four goes into twenty-eight seven times, so twenty-eight divided by four is seven. What does this *really* mean to you?"

"Well," I began, "it means twenty-eight divided by four is seven, because four goes into twenty-eight seven times." I really didn't think I was helping the situation here, but that was the best I could do.

The Hermit puzzled over my statement for a couple of minutes. He cocked his head one way, and then another. He smoothed the sand under his foot with his sandal. I wanted to ask what he was thinking, but I didn't have the nerve. Finally he stood up and stated that he had remembered something he had to do this morning, and trundled off to the rear of the cave. I had no idea if he was upset with me or I had simply triggered something with my statements that reminded him of a forgotten chore. After what seemed like half of the morning he finally returned. He was carrying a small, cotton sack with handles. He laid it at my feet and asked me to remove the contents. I did, and it contained what turned into a pile of AA batteries at my feet.

"In addition to my canning I wanted to check my other supplies today, to make sure that I was not running low. These are all fresh batteries for my graphing calculator. Would you see how many sets we have there."

"How many does the calculator take each time you replace the batteries?" I asked.

"It needs four at a time." He said. "I need to know how many replacement sets I have in that pile. You know, you never have enough batteries."

I began to sort out the batteries to determine how many sets he had here. I had hardly begun when I heard, "Stop! What are you doing?"

Division as a Repeated Subtraction

I almost fell off of my stool, he said it so loudly! I had no idea why he had stopped me, or what I was doing wrong. I did, however, have to answer.

"Well, I was simply trying to determine how many sets of batteries you had for your calculator. I thought I was doing it correctly."

"You were," he stated. "I didn't mean to frighten you, but I was really amazed at what, exactly, you were doing there with the batteries."

"Well, I took four and put them here." I pointed to four batteries to my far right. "Then, I took four more and put them here." Again, I pointed to a stack of four batteries next to the first stack.

"So, what I hear you saying is that you 'took four away from the large pile', and then continued to do so, repeatedly taking four from the initial pile."

"Yes, that's right. I was going to continue until there were none left there."

"So you were going to continue to *subtract* from the original pile until there were none left."

"I guess…."

"Good, continue your task. I want to see it completed."

I continued removing batteries four at a time and putting them into smaller piles. When I had finished I counted the small piles of four that I had constructed. I counted a total of seven of these piles.

"You have enough batteries to refresh your calculator seven times." I stated. I knew that it was a job well-done.

"And you found this out by taking away four, again taking away four, etc., until there were none in the original pile?" He asked.

"That would be correct." I answered.

"So in order to *divide-up* the batteries, you repeatedly *subtracted*?" He said in a questioning tone.

"Yes, I guess I did." My reply was hesitant.

"What exactly do you think that you are telling me here?" Once again, a question that needed an answer from me.

"What this must really mean is that in order to *really* perform a division, when we divide something up in the real world, we have to repeatedly take away the same amount from some 'pile' of things. This reminds me of …"

"… our definition of multiplication."

"Yes! That was a repeated addition of the same number. Now division is simply a repeated subtraction of the same number."

"Excellent!" He said this excitedly. "Just as we found multiplication to be a short-cut for a repeated addition, we now have division as a short-cut for a repeated subtraction.

Division as Separation into Groups

"Well, that was a good mental work-out." I stated. "Now I know the 'real' definition of division."

"Simply stating that division is a short-cut for a repeated subtraction is nice, but what we need here is to ascertain the meaning of the results. I need more liquid refreshment before we begin this task. I see your glass is almost empty also. Let me refill these before we continue. What you can do is determine how many oatmeal cookies are left for each of us. I'll be right back." With that he picked up the two glasses and disappeared into the depths of the cave.

When the Hermit returned he saw that I had separated the cookies into two piles, one for him and one for me. He asked how I had done it, and I answered that I simply moved one to his side of the plate, then one to mine. I repeated this until there was a pile of seven on his side and a pile of seven on mine.

"So," he said, "you subtracted only one each time in this excerise."

What, Exactly, is Happening Here?

There was quiet for a couple of minutes as the Hermit thought about exactly what had happened here. I didn't know if I had done something wrong or he was simply contemplating what to do next. Finally he spoke.

"I see that you first subtracted four at a time from the group you had in front of you, and then, the second time, subtracted only one at a time. Why, exactly, did you do these two exercises this way?"

I thought about this for a short time. I did not even realize that I had done the problems two different ways. I finally recognized the difference.

"Well, in the first problem I *knew* how many batteries were in each group, so I simply put them in groups of four. In the second problem I *did not know* how many cookies were to be in each group, but I *did know* how many groups there were. I placed one in each group, one at a time, until there were no more cookies in the original batch."

He responded, "So, I assume that to perform a division as you have been doing you either need to know how many are in each group or how many groups there are."

"I guess that is correct."

"Well, then, let's examine another example. Let's first assume that we have a pile of objects (and the pile contains 35 objects, but we will pretend we do not know that). Now, let us further assume that we want to divide the pile into groups of five. We would have the following division problem:"

He then proceeded to pick up his stick and write in the sand:

$$35 \div 5 = 7$$

"This means the following," he stated. "If we have a 35 objects and divide them into groups of five, we will end up with seven groups. It would look like this:"

35	÷ 5	= 7
Thirty five objects	put into groups of five	yields seven groups.

"or … if we have 35 objects and divide them into groups of 7, we will have 5 groups:"

35	÷ 7	= 5
Thirty five objects	put into groups of seven	yields five groups.

"So," I said, "this is the same problem, only we are looking at it from two different perspectives!"

"That is exactly right. Many people perform the operation of division without even realizing that they may be looking at one of two different problems. But … we have yet to tackle one important part of division, and that is the actual operation itself. Once you *know* what the operation of division means, and hopefully by now *you do know* this, the next question is how to perform the operation on larger numbers!"

Multiply, Subtract, "Bring down"—This is Division?

A heavy quiet fell over us for a short time, then the Hermit spoke.

"We do not always want to separate a number of objects into groups, and this is for a couple of reasons. One reason is that it becomes cumbersome when the numbers we are dealing with are large. Another reason is that we may not always have the objects close at hand. So, we need another way to handle the problems when these instances occur."

I was ready. I said confidently, "I know, we can use the process of long division. We do not need the objects here, and it does not matter if the numbers are large, the process is always the same."

"You are most certainly correct. Maybe I could give you a problem now and you can demonstrate the process of 'long division.'"

"Certainly. I would be happy to show you how I do it."

"Good, here is the problem. I want you to divide 6516 by 12. That would seem to be a good place to start."

"I will do the problem step-by-step so that you can see what I am doing. I will explain each step so that you can easily follow."

```
         5                    5                   54                  543
    12)6516              12)6516             12)6516             12)6516
      60                   60                  60                  60
                           51                  51                  51
                                               48                  48
                                                                   36
                                                                   36
                                                                    0
```

⬆ First I have to divide 12 into 65. Then I multiply 5 and 12 – the answer is 60.

⬆ Now I subtract, the remainder is 5 and I will bring down the 1.

⬆ Twelve goes into 51 4 times, and 4 times 12 is 48.

⬆ Now I subtract 48 from 51 – the answer is 3. I bring down the 6 and 12 divides into 36 exactly 3 times!!

"There are some questions I have in your process," stated the Hermit. "It seems as if you were simply 'guessing' when you said that 12 divides into 65 and 12 divides into 51, etc. In reality, 12 divides into *neither* of those evenly! Were you really *guessing* in order to do a mathematics problem? My second question is the following: Why is it that you do this 'bring-down' thing? Is that a real mathematical operation?"

I was at a loss here. This was *always* the way I did division. I had to say something, but I was truly at a loss for words.

"Well," I started, "that is simply the way you have to do division. I suppose that sometimes everyone has to guess when doing mathematics."

"You realize that you were just following steps in a process without knowing why you were doing them. You were reducing mathematics to a rote process that has very little, if any meaning. Remember, what you do in mathematics should *have meaning* and *be logical*. If it does not have these properties, then you are simply reducing mathematics to a memorization of a lot of rules—rules that have no meaning in the real world. Your answer is correct, but we should be able to determine *why* it is correct."

"This is great. I would really like it if you would show me why my process is correct."

"Good, I think you are ready."

THIS is Division—with Meaning

"Let's look at your problem one more time. We want to divide 6516 by 12. Obviously we do not know the answer or we would not go through all of the trouble of performing the process of long division. Let's look at your problem again. It looked like this at the beginning:

```
12)6516
```

"The question says 'how many times can 12 go into 6516.' I am going to 'guess', much like you did. I am going to guess that it divides 500 times, even though I know that I am wrong."

"But, ... but ... you said guessing was wrong when I did it."

"I never said that it was wrong, but said that you did not know why you were doing it. Sometimes an intelligent guess in mathematics can be useful. We often use this method in *estimations*. So ... *my estimate* for the division is 500. Now the process looks like this:

$$\begin{array}{r} 500 \\ 12\overline{)6516} \end{array}$$

"When I multiply 500 and 12 I find that my estimate was close, but not correct. Now I will need to adjust it.

$$\begin{array}{r} 500 \\ 12\overline{)6516} \\ \underline{6000} \\ 516 \end{array}$$

"I can now see that my estimate was incorrect by 516. Now I need to estimate how many times 12 divides into 516. If it were 500 ÷ 10 the answer would be 50. I will try 50 as my refinement of my original estimate.

$$\begin{array}{r} 550 \\ 12\overline{)6516} \\ \underline{6600} \\ 516 \\ \underline{600} \end{array}$$

"Now, and this sometimes happens when you 'estimate', my estimate of 50 is too large because I cannot subtract 516 – 600!! So ... I will try revising my estimate from 50 to 40. Now my division looks like this:"

$$\begin{array}{r} 540 \\ 12\overline{)6516} \\ \underline{6600} \\ 516 \\ \underline{480} \\ 36 \end{array}$$

"Notice that I was now able to subtract 516 – 480, the result is 36. I can easily see that 36 ÷ 12 = 3, and I can finish the problem."

$$\begin{array}{r} 54{\scriptstyle3} \\ 12\overline{)6516} \\ \underline{6600} \\ 516 \\ \underline{480} \\ 36 \\ \underline{36} \\ 0 \end{array}$$

"Long division is simply a process of *estimation and revising your estimates*. As your estimation skills become better you will be able to increase the speed at which you divide."

Session 5—The Order of Operations

Introduction

I recall very vividly one particular day when I visited with the Hermit. This was a beautiful balmy day, and we were sitting in the sun in front of the cave. For some time we sat saying nothing, simply enjoying the breeze and knowing that this was going to be a beautiful spring. I was sitting and watching the clouds. The Hermit was doodling with his graphing calculator, saying nothing, but pressing keys, clearing the display, and pressing more keys.

I finally had to ask him what he was so curious about. He didn't say anything for a couple of minutes, simply content to continue entering into and clearing the display. Finally he said, "This would make an excellent tool for the teaching of the Order of Operations."

"The Order of Operations!" I repeated, startled at his sudden statement.

"Yes, it does quite well in doing the mathematics correctly."

"But, doesn't everyone already know the Order of Operations?"

"Ah, you tell me. Yes, right now, you tell me exactly what you think the Order of Operations is."

"Well, first you simplify everything inside the parentheses. Secondly, you simplify any exponents. Next you do multiplications and divisions. Lastly you do the additions and subtractions."

"And that's it? It sounds so simple."

"It is. It really is." I could tell that he was really impressed by how simple and concise I had made it. I knew he was impressed.

Please Excuse My Dear Aunt Sally—for being Incorrect!

"There's even a very simple way of remembering it." I continued. "There is this cute little saying that makes this all so simple. It goes like this."

I picked up a small stick and wrote a column of letters in the sand.

P
E
M
D
A
S

I then said, "You can remember these letters with the sentence, 'Please Excuse My Dear Aunt Sally.' Each letter represents one of the parts of the Order of Operations." I then proceeded to finish each of the words that I had begun.

Please
Excuse
My
Dear
Aunt
Sally

"If you remember this, then you simply replace the words in the phrase with the words in the Order of operations.

Please (Parentheses)
Excuse (Exponents)
My (Multiplication)
Dear (Division)
Aunt (Addition)
Sally (Subtraction)

"See," I said, "I can't make it much simpler than this."

He thought about this for a few seconds; then, he scribbled a few symbols in the sand. This is what he wrote:

$$\frac{3+4}{7}$$

"That's simple," I replied, "You simply add the 3 and 4, then divide by 7!"

"Aha! You just violated your own rule. This expression includes two operations: addition and division. Yet … you added first when you should have divided!"

"Well, that expression is *different*. You *have to* add first."

"Let's look at another simplification. This one will be a simple one to simplify—just apply your 'PEMDAS' rule. Here it is:"

He proceeded to clear the sand and draw the following:

$$8 \div (4)(2)$$

He than stated, "You, according to 'Aunt Sally', would multiply first and then divide. You would find that the following is the answer:"

$$8 \div (4)(2) = 8 \div 8 = 1$$

He continued without any interruption, "Or … you can '*do the parentheses*' first, then continue. The result is the same."

$$8 \div (4)(2) = 8 \div 8 = 1$$

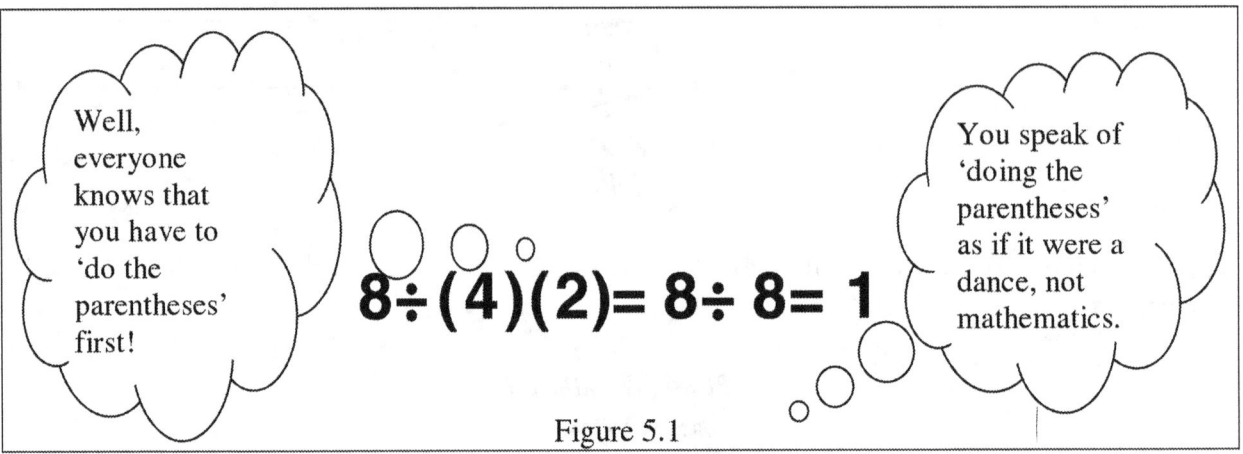

Figure 5.1

"This result is equally incorrect!"

"Why is this?" I inquired.

"Because," he said, "the Order of Operations states that you simplify *inside the parentheses* first, then you do the operations of multiplication and division *from left to right as they occur.* Finally, you perform the operations of addition and subtraction *from left to right as they occur.* This is the *real* Order of Operations. *What most people forget are the important words 'inside the parentheses' and 'from left to right.'*

"Your little mnemonic is cute, but most people only tend to remember the six words in the saying, 'Please excuse my dear aunt Sally.' The importance in this is the words that they do not remember, the ones that make the process work!

"In this problem you should realize that there are two operations, multiplication and division. The parentheses simply indicate a multiplication. Reading from left to right, the division should be performed first, then the multiplication. Hence, here is the **correct** simplification:"

$$8 \div (4)(2) = 2(2) = 4$$

"So, My Dear Aunt Sally is ..."

"Wrong, and for a variety of reasons. We have just begun to scratch the surface."

"Well, what is a good way of remembering—a way that will help to eliminate errors?"

"That's where we are going right now," he stated as he stood up to stretch. "First, let's have some lunch, and then we will continue this conversation on a full stomach."

The REAL Order of Operations

After our lunch it was becoming a bit warm sitting in front of the cave opening, so we decided to sit just inside the entrance. There was plenty of light from the sun, but the shade of the cave and the breeze kept us cool. The Hermit was once again fiddling with his calculator as I cleaned up the remnants of our meal. He could play the keys on that thing like Liberace could play the keys on a piano. Finally I finished my chore and sat on a stool near him. I really wanted to know about the Order of Operations ala the Hermit, but waited patiently. Finally he spoke,

"How many operations do you know in mathematics?"

"Well, there is addition, and subtraction, and multiplication, and finally division."

We had been down this road already today, and I was becoming very curious as to which fork we would take this time.

"And exponents?" he asked.

"Well, yeah, I guess that is an operation. Sure, I can agree to that."

"Let's now just consider addition and subtraction. Do you recall how we defined them some time back?"

"Sure, with both you need a pile or group of things. For addition you put more into the group, and for subtraction you remove some from the group."

"OK, let's look at a problem with additions and subtractions. Tell me what you see."

He picked up a stick and wrote the following on the floor in front of his stool. Amazingly, he wrote the whole problem upside down so that it was right side up facing me. Here is what he wrote:

$$26 - 10 + 7 - 12 - 4 + 9$$

"In terms of your definition, what does this mean to you?" he asked.

"Well," I hesitated, "I guess we start out with 26, then remove 10 from the group, then replace 7, then successively remove 12 and then 4, finally replace 9 into the group."

"So … in terms of the symbols, we can simply read and perform the operations from left to right, just as we read from left to right."

"Sure, and the answer is 16."

"Exactly. Now, do you remember how we defined multiplication?"

"Yes, we defined it as a repeated addition—as a short-cut for multiple additions. We said that 4 x 5 = 5 + 5 + 5 + 5. You said that 4 x 5 simply means *4 times add 5*."

He began smoothing the dirt in front of himself, preparing to produce another example. He wrote the following:

$$10 + 3 \times 4$$

He began to explain what he saw on the ground, "What I see here, then, is nothing more than an addition problem. I could easily write the following:"

$$10 + 4 + 4 + 4$$

"What you have to understand is that this is all addition. Any problem involving multiplications is just an addition problem."

I was stunned! I had to ask, "So … there is no such thing as multiplication?"

"Well … yes and no," he replied, "but it is simply a shorter way of writing an addition. If I have a problem containing 23 x 9, I certainly do not want to write twenty-three nines and then add them. I have this nice short-cut called multiplication that gives the answer of 207. But … I also have to realize that *this is really just an addition problem*. Suppose I have the following problem:

$$6 \times 9 + 3 \times 4$$

"I know that this is all addition. I have to add six nines, and then add to that three fours. My answer looks like this

$$6 \times 9 + 3 \times 4 = 54 + 12 = 66$$

"And … if I have this problem

$$6 \times 9 - 3 \times 4$$

"I *still* have to see what the 'short-cut' of multiplication gives before I can even begin to consider the subtraction."

$$6 \times 9 - 3 \times 4 = 54 - 12 = 42$$

"So," I said hesitantly, "I have to do the short-cut for addition, called multiplication, before I do the additions and subtractions. Correct?"

"Precisely. Now let's examine division. Do you recall our definition of division?"

"Of course, we said that division was no more than a repeated subtraction, or a short-cut for doing multiple subtractions of the same number. We said that if we had a pile of pebbles and wanted to divide the pile by 4, we simply remove 4, then remove 4 again, and continue the process until none are left. We then counted the number of piles of 4 we had. That was the answer."

As soon as I finished I realized what I had just said. The Hermit must have recognized it also, as he was sitting there with a wry grin on his face.

I hesitated, and then I continued, "Division, then is just like multiplication. It is nothing more than a short-cut for repeated subtraction. If I encounter a division problem within a larger problem, I have to realize that this is just a subtraction problem. I need to find the result of that subtraction before I can continue."

"Precisely. The short-cuts of multiplication and division must be done first. Let's see how you would handle this problem." Once again he smoothed the sand and wrote (upside-down) another problem to be simplified.

$$(3)(5) + 6 \div 2$$

I muttered quietly, almost to myself, but wanting him to hear, "I know that this is all simply addition and subtraction. I first need to know what the short-cuts of multiplication and division give, then add the results." I wrote the following:

$$(3)(5) + 6 \div 2 = 15 + 3 = 18$$

"Good job," he said, "Now let's try another. We'll clear the slate and see how you do on this one."

$$12 \div (2)(3)$$

"Ok," again muttering quietly to myself, "I know that there is nothing more here than additions and subtractions. It seemed that I had been 'tricked' into the incorrect answer earlier in the day. I read it from left to right. 12 divided by 2 is 6, and I then have 6 times 3. 6 times 3 is 18." I proudly wrote the following:

$$\underbrace{12 \div (2)(3)}$$

$$= \quad 6 \quad (3)$$

$$= \quad 18$$

"Excellent," he said.

"So, that's it! We're finished?" I asked.

"No, not yet. We have another operation. We have yet to deal with exponents. We must find out how these fit into our scheme. "You *do* recall the definition of an exponent, do you not?"

"Of course I do," I stated proudly. "An exponent represents a repeated multiplication of the same number. 3^4 simply means 3 x 3 x 3 x 3, which is 81."

As soon as the words were out of my mouth I had realized what I had just said. My jaw simply went slack but my brain was working diligently. I was *sure* where this was going.

"So," I began, "an exponent is simply a short-cut for repeated multiplications, which, in turn, are no more than short-cuts for repeated additions. An exponent is nothing more than a short-cut of a short-cut!"

"And in order to perform a simplification ..." he left the answer hanging.

"you first have to determine the value of the exponential part, which is a short-cut of a short-cut, then perform all multiplications and divisions, and finally you can add and subtract. Let's see if I can do an example correctly," I asked.

"OK, here is a nice one for you," he stated as he once again smoothed the sand.

$$8 \div 4 + 3^2 - 2 \times 3$$

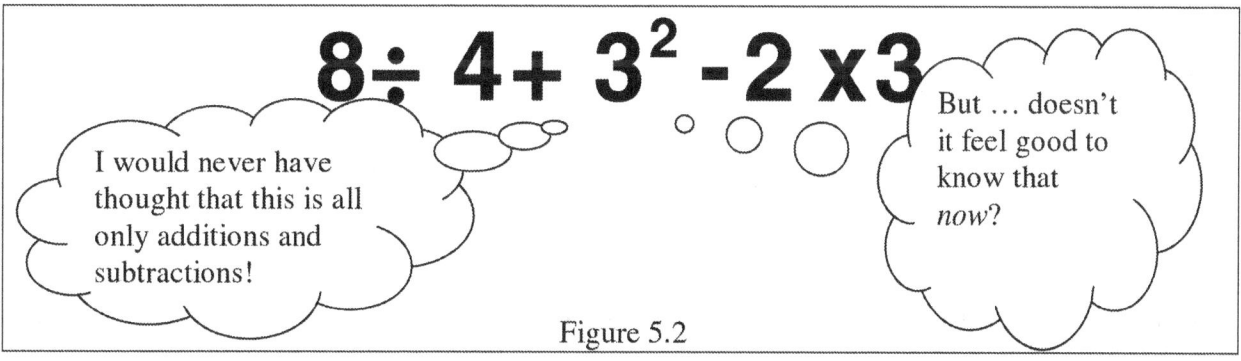

Figure 5.2

"Well, first I am going to simplify the 3^2, then simplify the division and multiplication portions," I said as I wrote another line.

$$8 \div 4 + 3^2 - 2 \times 3$$
$$= 2 + 9 - 6$$

"And my final answer is 5!"

"Very good. You are learning a lot today," he stated. "But, now it should be my turn. I would like to do one. You can watch."

$$7 - 4 + 2 = 7 - 6 = 1$$

"No, no," I exclaimed, "You did the problem incorrectly. You should have done the subtraction first, then the addition. You should have gotten $7 - 4 + 2 = 3 + 2 = 5$!"

"You are, indeed, correct. But … what if I wanted to add first, and then subtract. I should have a method of letting you know that the answer is really *one*. There should be a way of *overriding* the Order of Operations as we have discussed it. We can do this by inserting grouping symbols."

"You mean parentheses?"

"Parentheses are just one type of grouping symbol. There are many others in mathematics. I could write my expression like this:

$$7 - (4 + 2)$$

"or this

$$7 - [4 + 2]$$

"or, like this:

$$7 - \{4 + 2\}$$

"All of these symbols, called **parentheses**, **brackets**, and **braces**, can be used interchangeably."

"So there are three different kinds of grouping symbols?"

"Actually, there are a number of different ones we have not even discussed. Would you like to guess at some of them?"

"*Actually, no.* But … I would like to know of the others."

"OK. Let me show you some others. I will even give an example of each to give you a good idea of what is correct and incorrect when using these symbols."

"First, there is the **absolute value** symbol. An expression using this would look like this:"

$$| \, 6 - 4 \, |$$

"The incorrect simplification would be if you first took the absolute value of each number, then simplified."

$$| \, 6 - 4 \, | = 6 + 4 = 10 \leftarrow incorrect$$

$$| \, 6 - 4 \, | = | \, 2 \, | = 2 \leftarrow correct$$

"There is the **radical**, or **square root** symbol. When you encounter this symbol you also *simplify inside the grouping symbol before continuing*."

$$\sqrt{25 - 9} = 5 - 3 = 2 \leftarrow incorrect$$

$$\sqrt{25 - 9} = \sqrt{16} = 4 \leftarrow correct$$

"There is also another symbol that many people take for granted, using it correctly and not even recognizing that it is a grouping symbol. It is the **fraction bar (or vinculum)** that we encountered earlier in a simplification. Our earlier example looked like this:"

$$\frac{3 + 4}{7}$$

"Since the horizontal fraction bar is a grouping symbol, we first simplify the numerator (if possible), then the denominator (if possible), and lastly we divide:

$$\frac{3 + 4}{7} = \frac{7}{7} = 1$$

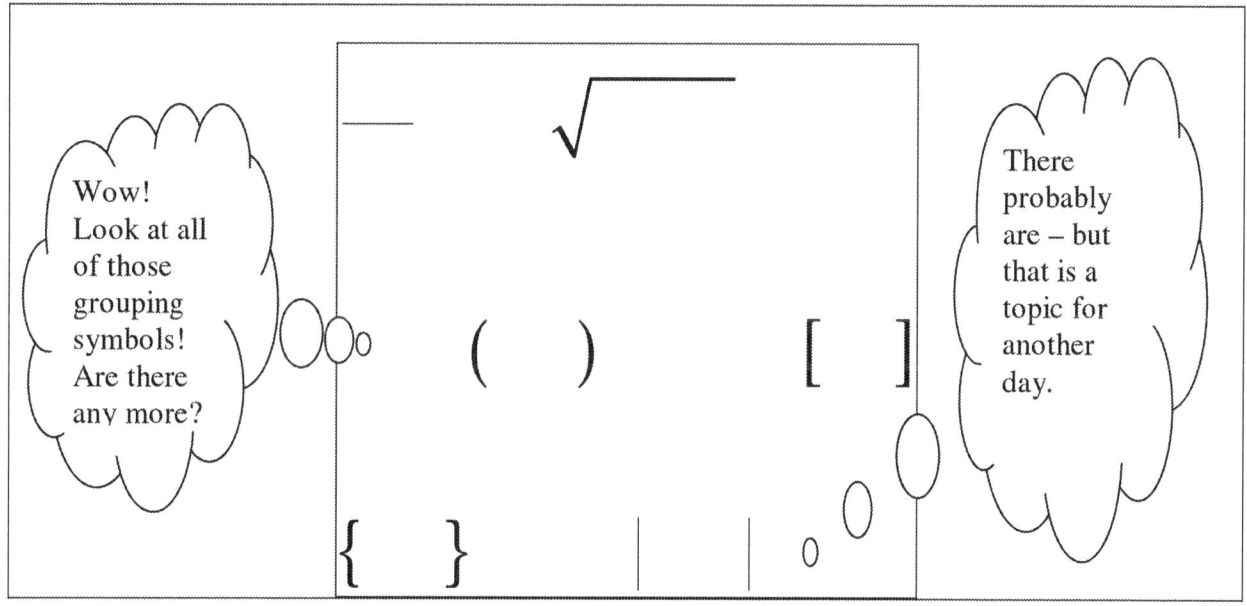

Figure 5.3

"Wow!" I said, startled. "Most people perform this simplification without even knowing that they are actually using the Order of Operations, and doing so correctly!"

"That's so correct. You will usually be told something like, 'Well, that one is different. It doesn't need the Order of Operations.' Are there any other questions regarding the Order of Operations floating around in your head?"

"Well," I hesitated. "I was experimenting with your calculator earlier and I keyed in the following."
I picked up his calculator from its resting place on the small stool-sized rock and keyed in this expression:

$$9 + 6 \div 3$$

"But," I continued, "the calculator screen showed the following:"

$$9 + 6/3$$

"and … when the calculator simplified the expression it did not treat the division symbol as a grouping symbol, but divided first and the answer was 11!"

"Ah, my young friend, you have indeed made an interesting discovery. It is also a discovery that makes many an algebra student want to pull out all of their hair. You have in your hand one of the miracles of modern technology, the graphing calculator. But, it also has a modern short-coming. It cannot display

$$\frac{9 + 6}{3}$$

"Because it cannot display two lines at once. The only division symbols are the one on the keyboard, '÷', and the one it displays, '/'. We stated earlier that the *horizontal* fraction bar is a grouping symbol. Unfortunately, neither of these which you just introduced is a grouping symbol! In order to show your problem with grouping symbols, you will need parentheses in the expression. If you try keying in this expression

$$(9 + 6)/3$$

"you will see that you will get the answer that you want. We will have to take some time one evening, break out the calculators and some hot cocoa, and see precisely what these tools can and cannot do!"

"Before we take a break from all of these symbols, let me show you a couple of other things while we are here. Do you recall the relationship between multiplication and division that allows you to perform one or the other operation in a problem?"

"Why, yes. Instead of dividing you can multiply by the reciprocal."

"Correct. So, watch my simplification change here:"

$$\frac{3 + 4}{7}$$
$$= \frac{1}{7}(3 + 4)$$

"Can you agree that dividing by 7 is the same as multiplying by $\frac{1}{7}$?"

"Yes, of course."

"Then we come to the following simplification:"

$$\frac{3+4}{7}$$

$$= \frac{1}{7}(3+4)$$

$$= \frac{1}{7}(7)$$

$$= 1$$

"Wow, I never thought of that," I said, startled. "I like that simplification. And … it's so logical."

"Mathematics *is* logical!" He stated with a small sneer in his voice. "If you recall the distributive property you could also do this:"

$$\frac{3+4}{7}$$

$$= \frac{1}{7}(3+4)$$

$$= \frac{1}{7}(3) + \frac{1}{7}(4)$$

$$= \frac{3}{7} + \frac{4}{7}$$

$$= \frac{7}{7}$$

$$= 1$$

I had to ask him the question that was floating around in my mind, but was afraid of the answer. I did, however need to know. I finally blurted it out,"

"But … how do I know which way to do the problem?"

He scratched his head while looking at all of the things that were drawn there in the sand. He finally answered slowly,

"There are but a few rules to follow in mathematics. As long as you follow the rules you know to be correct, your answer should also be correct. You simply do what suits you best – what gives you the best chance of doing the simplification correctly. *Mathematics is logical.* You simply have to be logical also."

Hermit Note: I explained the Order of Operations to my friend in terms of the numbers of arithmetic, or the natural numbers. Had he been more knowledgeable in the concepts of algebra and integers I would have proceeded differently. Those of us who are acquainted with algebra know that you almost never see any rule for the subtraction of integers. The only rule stated for this subtraction will look something like this:

In order to subtract an integer b from an integer a, do the following:
a – b = a + (-b)

In other words, what I would have told my friend is that not only is there no multiplication, since it is simply a short-cut for performing addition, and that there is no division, but it is simply a short-cut for subtraction, but that there is not an operation of subtraction, it is simply an addition of signed numbers.

The **real** Order of Operations is based on the fact that the only **real** operation is addition. Subtractions are simply additions of the opposite. Here is one more simplification for you to examine. I have to run, I have a casserole in the oven and don't want it to burn.

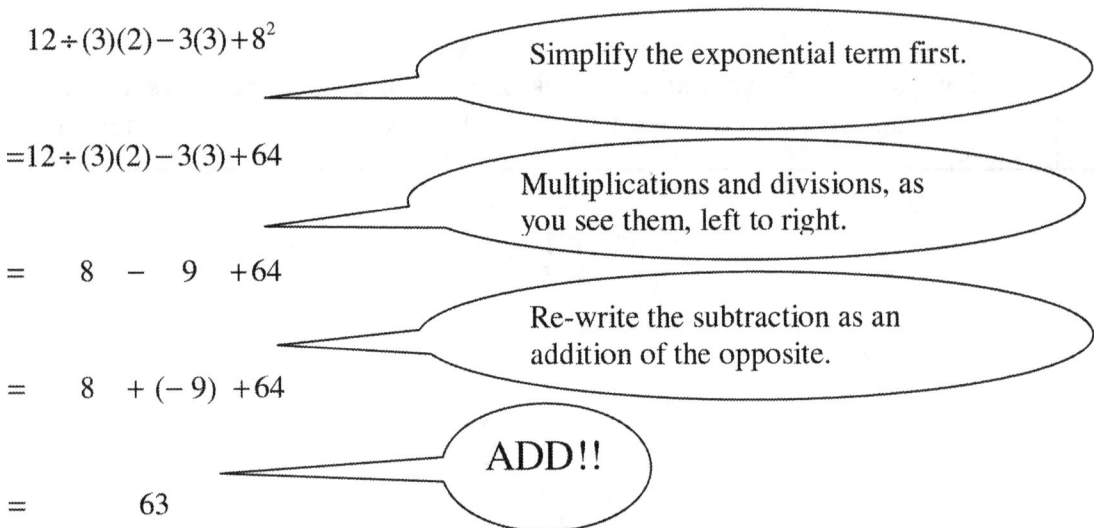

$$12 \div (3)(2) - 3(3) + 8^2$$

Simplify the exponential term first.

$$= 12 \div (3)(2) - 3(3) + 64$$

Multiplications and divisions, as you see them, left to right.

$$= \quad 8 \quad - \quad 9 \quad + 64$$

Re-write the subtraction as an addition of the opposite.

$$= \quad 8 \quad + (-9) \quad + 64$$

ADD!!

$$= \quad 63$$

Session 6—Fractions

Introduction

"I have a question for you, if I may." I said this hesitatingly, not really wanting to break the peaceful silence there in the Hermit Cave.

I was being ignored, but not intentionally. The Hermit was busy replacing the batteries in his graphics calculator. I had brought him fresh ones and he was excited.

"You know, I always like to have fresh batteries in my calculator. The colors of the display are so much nicer with fresh batteries."

"I still have a question," I said, rather petulantly.

He put the calculator in his lap and simply looked up at me. Nothing was said, so I assumed that it was time for me to speak and for him to listen.

"Why are fractions so difficult?"

"I didn't know that they *were* difficult!"

"Well, nobody likes fractions."

"I like fractions. I think that they are fascinating."

"Well, I guess I meant *normal* people!! But really, do we need all of this fraction stuff?"

"I couldn't eat a half dozen donuts without fractions. I couldn't buy a half gallon of ice cream … or a quart of milk. I couldn't cut a recipe in half and make only a small quiche. I couldn't share a pizza with three friends so we would each have one-fourth of the pizza.…"

"OK, you win. We *need* fractions!"

"Now that I answered your question, I have one for you. What, exactly, do you think that these things called fractions really are?"

The Meaning of Fractions

I had the answer to this one ready. This was one time I was going to get the answer right, and on the first try.

"A fraction is part of a whole. To follow up on your previous examples, a slice of pizza is a fraction of the pizza. A half dozen donuts is a part of a box of donuts."

He took this in, listening without speaking. He was pondering his answer while running his fingers over the keys of his calculator. He almost seemed to be petting it as if it were a cat on his lap. He reverently placed it on the small table to his left, picked up a stick, and decided to draw in the sand in front of his stool.

"Let's look at a fraction—tell me how this fits your definition."

He wrote the following in the sand:

$$\frac{3}{4}$$

"How does this fraction fit your definition?"

"Well three-fourths is three-fourths of *one*!"

"OK, let's look at another." Next to the three-fourths he wrote the following:

$$\frac{7}{4}$$

"That's an improper fraction. It is larger than one."

"Let's stay with your definition. I want to know how seven-fourths fits your definition."

"Well ... seven-fourths is a part of two!" I had him here, and I could feel it.

"Is not *one* a part of *two*? Yet, *one* is not a fraction, but a whole number itself."

I could see that I was not going to get very far in this direction. I would have to give a nice, precise, and mathematical definition of a fraction in order to continue this conversation.

"OK," I started, "a fraction is nothing more than a division problem in which we give the parts fancy names. In the fraction *three-fourths* we call the *three* the numerator and the *four* the denominator. But ... it is still just a division problem."

"What you say is good ... up to a point. But a fraction is much more than *just a division problem*. When I say that I have eaten six of the dozen donuts in a box, I do not think in terms of division. If I see eight slices of pizza in a box, and I proceed to eat three, I don't think in terms of division. There are so many ways in which we use fractions without thinking about it being a division problem." He stood up and reached into his pocket, he pulled out a quarter. "This is one-fourth of a dollar. But ... I don't think of division, but of what I can purchase with it. Should we get on with the useful definition of a fraction?"

My head was spinning and I wasn't sure where we were going. But ... I did want to know more.

I jumped right in, "Yes, let's see this *useful* definition."

"Let me draw it," he said as he picked up his stick. He wrote the following in the sand.

$$\frac{\text{the number of equal pieces you have}}{\text{the number of equal pieces in a unit}}$$

He continued, pulling two more coins out of his pocket. "I now have three quarters. They are all of equal size. If I had four, I would have a dollar (call it *the unit*). Thus, I have three-fourths of a dollar.

"Now, look at this," he said as he laid down two of the quarters. "What do you see in my hand now?"

"I see you are only holding a quarter, which means you have one fourth of a dollar."

"Or ... I have *one quarter*, or *one unit*, which has in it *twenty five* equal pieces, each called a cent! Or ... I have *one quarter*, or *one unit*, which has in it *five* equal pieces, each called a nickel! This is what makes my definition of a fraction so useful and meaningful. You define your unit and determine how many equal pieces it can be divided into. For instance, let us look at our other example." He dug into his pocket before continuing. Before I knew it there were *seven* quarters in his hand.

"Recall our example of the fraction seven fourths. I now have seven equal parts in my hand, four of which make up the unit called a dollar. Therefore I have one dollar and three-fourths of another dollar. I have one and three fourths-dollars. What I am holding, in terms of fractions, is dependent upon how I define my unit. For instance, what if I took a five-dollar bill out of my pocket, and decided that this was my *unit*. What fraction of that would I now have?"

"Well, each dollar has four quarters, and five dollars would contain five times four, or twenty quarters. You would be holding seven-twentieths of five dollars."

"Good! Now let's really test those gears between your ears. If five dollars contains 500 cents, what fraction does my seven quarters now represent?"

"I need to write this one out. I'm pretty sure I can do it if I can see it." I began to write in the sand.

$$\textit{five dollars} = \textbf{500 cents}$$
$$\textit{one quarter} = \textbf{25 cents}$$
$$\textit{seven quarters} = \textbf{175 cents}$$

answer: $\dfrac{175}{500}$ **(what I have)** / **(in one unit)**

"Good, let's take a break, maybe get a small snack and a walk."

Reducing Fractions, or Putting into 'Lowest Terms'

Our 'small' snack consisted of finishing the box of donuts!! We didn't discuss fractions or numbers, simply made the donuts disappear. I had, however, been staring into my scratching in the sand. Something about my answer was bothering me, but didn't seem to affect the Hermit. Finally I had to blurt it out.

"My answer is not in lowest terms. How could you accept that as being correct? Shouldn't fractions always be in lowest terms?"

"Actually, you found what I asked you to find. I asked you what fraction of 500 cents my seven quarters represented. I wanted the answer as a fraction with a denominator of 500. But, here is an interesting observation. Watch:

$$\frac{175}{500} \text{ (in cents)} = \frac{7}{20} \text{ (in quarters)}$$

"Notice that the amount of money did not change, simply how we looked at it. But, I *am* curious. I would like to know what you mean by putting a fraction into 'lowest terms' really means. I, like you, often hear people say 'lowest terms' or 'simplest form', and am curious, really curious, about what, exactly, they *think* they mean."

Explaining this was not going to be easy. I knew what I meant by reducing fractions, but it was not an easy thing to explain. I decided that the best thing I could do was to offer an example, to show the Hermit what reducing fractions really looked like mathematically. I stated, after thinking about this, "Here, I will show *you* an example that demonstrates how to reduce a fraction. "Here is the example," I stated as I drew in the sand. This is what I wrote:

$$\frac{54}{72}$$

"There," I stated, "now I will reduce this to lowest terms." I wrote the following:

3

~~9~~

~~27~~

$$\frac{~~54~~}{~~72~~} = \frac{3}{4}$$

~~36~~

~~12~~

4

I was proud of my simplification of the fraction. I said so, "There is the way you reduce fractions."

"Interesting," was all he said. He continued, "Let's take a break now, I see a flock of migrating Sandhill Cranes in the distance. We can watch them fly by us."

Reducing Fractions, or Putting into 'Lowest Terms', Part II

We returned from watching a flock of migrating cranes fly over our. It was truly an amazing sight, seeing all of those birds flying overhead, there must have been over a hundred. Some were in lines, following each other; others were flying in the *vee* formation that the birds are famous for. We had another small snack (leftover quiche); then, the Hermit returned to my problem.

"You have a nice simplification of the fraction written there. Let me try another route, see if I get the same answer."

He then proceeded to write his simplification in the sand.

$$\frac{54}{72} =$$

13

~~26~~

$$\frac{~~54~~}{~~72~~} = \frac{13}{18}$$

~~36~~

18

"Here is my reduction," he stated proudly. He continued, "It looks as if we have a difference of opinion. Let's look at our solutions side-by-side." He then compared both of our solutions. It looked like this:

$$
\begin{array}{l}
3 \\
\cancel{9} \\
\cancel{27} \\
\cancel{54} \\
\hline \cancel{72} \\
\cancel{36} \\
\cancel{12} \\
4
\end{array} = \frac{3}{4}
\qquad\qquad
\begin{array}{l}
13 \\
\cancel{26} \\
\cancel{54} \\
\hline \cancel{72} \\
\cancel{36} \\
18
\end{array} = \frac{13}{18}
$$

"It looks as if one of us is wrong," he stated. "We need to find out if one or both of us made an error. Can you tell me what you did in your simplification?"

"Well ... er ... I think ... I can't remember."

"I can understand that. I can't even see the numbers that I started with!! This kind of simplification is all right if you simplify correctly. But ... if you make a mistake and need to find it, well, this is very difficult. Exactly what are you doing when you reduce the fraction?"

"Well, I am looking for something in common in the numerator and denominator, then I divide both by that number."

"So, you want to find out what the numerator and denominator have in common. You know that you can do this by factoring the numerator and denominator into prime numbers. You *do remember* prime numbers?"

"Yes, they are numbers that only divide evenly by themselves and *one*."

"Good, watch me factor each of the parts of the fraction." He continued to smooth the sand; then, drew the following:

$$54 = 2 \cdot 3 \cdot 3 \cdot 3$$
$$72 = 2 \cdot 2 \cdot 2 \cdot 3 \cdot 3$$

He proceeded to explain; "Now I know that every number divided by itself is equal to *one*. I can now write the simplification of the fraction as follows:

$$\frac{54}{72} = \frac{2\times 3\times 3\times 3}{2\times 2\times 2\times 3\times 3}$$

$$= \frac{\overset{1}{2}\times 3\times \overset{1}{3}\times \overset{1}{3}}{\underset{1}{2}\times 2\times 2\times \underset{1}{3}\times \underset{1}{3}} = \frac{3}{4}$$

"I can now see that I divided *two* by *two*, *three* by *three* twice, and the answer to each was *one*. I can *see* what I did, and *find any mistakes* that I may have made. You cannot do that with your simplification! What do you think about this?"

"I was correct!! But ... I couldn't show you this. Wait ..., *you knew you were wrong* and I was right. You tricked me into a better method of simplifying these fractions!"

"You are right again. You had to see that there was a better method of reducing fractions. I needed a way to show you this."

"Can you give me another example, one that I can do *your* way?"

"Sure, let me see you simplify this fraction." He proceeded to draw the following fraction in the sand, after smoothing all the sand meticulously.

$$\frac{36}{54}$$

"OK, here is my simplification." I wrote the following:

$$36 = 2 \times 2 \times 3 \times 3$$
$$54 = 2 \times 3 \times 3 \times 3$$

"Now I have the following:"

$$\frac{36}{54} = \frac{2 \times 2 \times 3 \times 3}{2 \times 3 \times 3 \times 3} = \frac{\overset{1}{2} \times 2 \times \overset{1}{3} \times \overset{1}{3}}{\underset{1}{2} \times 3 \times \underset{1}{3} \times \underset{1}{3}} = \frac{2}{3}$$

"Very nice," the Hermit said politely. "Now, let me ask just one more question. What exactly was it that you did in terms of reality? You said that $\frac{36}{54} = \frac{2}{3}$, and that I can see. What I guess I really want to know is how does this relate to our previous conversations, those involving cents, quarters, and dollars?"

I was stumped! "I really don't know. I know that I reduced it properly to smallest terms, but don't know what this means … I guess."

"Well then, it must be time to find out what 'reducing' *really* is!!"

Reducing Fractions—What's Behind it All

"We will look at another example to get us started. We'll make it a simple one because I will need some pictures. Here is where we will start. Do you agree with this simplification?"

He proceeded to clear an area in the sand, and then wrote the following:

$$\frac{9}{12} = \frac{3}{4}$$

"Yes, that is certainly a easy-enough simplification."

"Good, now let's see what it really looks like. Here is a picture of nine-twelfths."

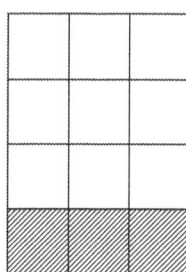

"Notice that my *unit* has twelve pieces. I have shaded the bottom three because I only want to look at nine, which is *what I have*. Since we know that both twelve and nine divide evenly by three, I can put these pieces into groups of three and there should be no remainder. Look:

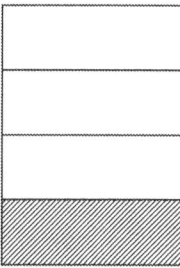

"By putting these into groups of three, I have created a unit with four pieces, of which I have three! I began with twelve *small* pieces in my unit, and now have four *large* pieces."

"So, I started, "you began with a lot of small pieces, and ended up with just a few large pieces. But ... the amount you had didn't change."

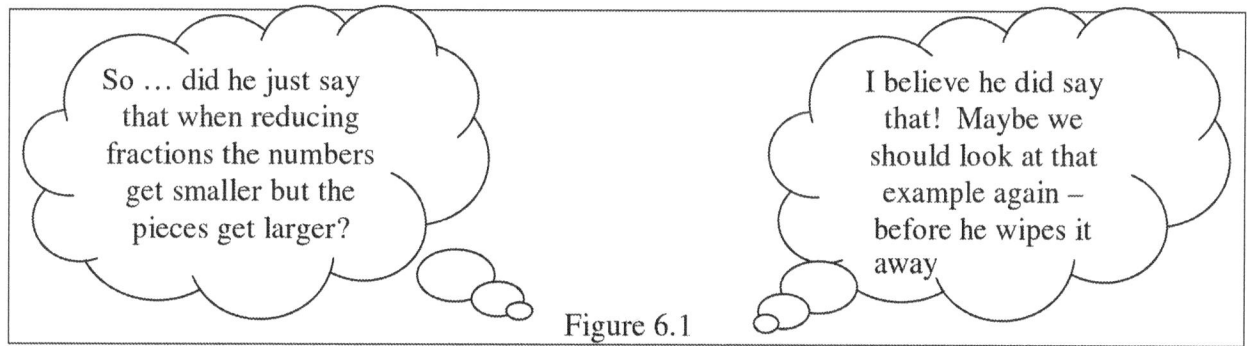

Figure 6.1

"Correct, but it's even better. When I re-grouped my original pieces by threes, I made my new pieces three times as large as originally. Let's look at one of our earlier examples. You recall the quarters and cents we discussed. We decided the following."

He cleared an area and wrote the following. I never could fathom how he remembered all of these things, and was able to duplicate them hours later.

$$\frac{175}{500} \text{ (in cents)} = \frac{7}{20} \text{ (in quarters)}$$

"You recall that we said that 175 cents was the same fraction as 7 quarters when the unit is five dollars. We can verify this by the following simplification:

$$\frac{175}{500} = \frac{7 \times \overset{1}{\cancel{25}}}{20 \times \underset{1}{\cancel{25}}} = \frac{7}{20}$$

"In order to change cents to quarters we simply make the size of the pieces in the numerator and denominator twenty-five times larger. I could also find out how many nickels there are in the same situation. How would I do this?"

"I would factor a five from both the numerator and denominator, making my pieces five times larger. Or … I am simply changing pennies into nickels!"

"Correct. I'll let you do this one."

I cleared a space in front of myself and wrote the following:

$$\frac{175}{500} = \frac{35 \times \overset{1}{\cancel{5}}}{100 \times \underset{1}{\cancel{5}}} = \frac{35}{100}$$

"There, I said proudly. 175 cents is $\frac{35}{100}$ of five dollars. In other words, there must be 35 nickels in the five dollars!"

"Precisely! Reducing fractions means nothing more than changing the size of the parts that you are interested in. When you reduce a fraction you are making a smaller number of larger pieces, but … the amount you have never changes.

Session 7—Fraction Multiplication

Introduction

I had awakened early this morning. This was the first time I had stayed overnight at the Hermit Cave. It was nice just sitting at the cave entrance watching the sky change from black to oranges and purples, then finally to blue as the sun broke over the cliffs. My mind drifted like the colors of the sky. Mostly it drifted to the conversations we had had the previous day. We had spent a large part of yesterday discussing fractions. I finally thought that I knew what fractions were, and was beginning to think that maybe they really were more useful than I had previously thought.

I had heard some stirrings in the depths of the cave, but was enjoying nature's sunrise too much to investigate. Eventually the silence of the emerging day was broken.

"Well, my friend," the Hermit said as he shattered the quiet, "I trust you slept well?"

I was briefly startled, but answered, "Yes, very nice sleep. I was just sitting here enjoying the sunrise."

"It's one of my hobbies. I often rise early just to enjoy the changing face of the land reflected in the changes in the sky. It only happens like this at dawn. But, I must admit, I needed to sleep-in today—our conversations of yesterday left me drained."

He was holding two mugs of steaming drinks in his hands. He held one out to me as he spoke.

"Thank you," I said before taking my first sip. It was hot, thick, and sweet. It tasted like a combination of coffee and cocoa.

"You have thought about our discussion of fractions?" He raised his voice at the end of the sentence to indicate this was a question.

"Yes," I replied, "and I think that I finally know what fractions really are."

"So, we can move forward with our discussion this morning?" Again, a questioning end to the sentence.

"Yes, by all means."

"You recall our definition of multiplication, do you not?"

"Sure, multiplication is a repeated addition. If I have the multiplication problem such as 3 x 6, it simply means add 6 three times. The answer is 18.

"So, then, what do you think that the multiplication of fractions means? If I have the following problem", he picked up a stick and began writing in the sand.

$$\frac{1}{2} \times \frac{2}{3}$$

"… this must mean that I add 2/3 to itself ½ of a time? That is correct, according to your definition.…"

"Well … that doesn't really make any sense, does it?"

"Not really. If we look at your original example of 3 x 6 a little differently, maybe we can transfer the result to fractions. Here is a picture of 3 x 6."

"I have drawn three rows of six items. There are a total of 18 items here. This is a nice geometric way of showing your 'repeated addition.' Three *times* I drew a row of six."

"I like that. It represents my repeated addition *and* I can really see it!"

The Geometric Meaning of Fraction Multiplication

"In order to do this we are going to have to draw some fractions like I drew the problem 3 x 6."
He smoothed the sand and began to draw.

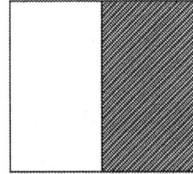

He then offered an explanation of his simple drawing, "Here is a representation of ½. If the whole square represents a unit, then I have shaded exactly ½ of it. This could represent ½ of a pizza, or even better, ½ of a pan of lasagna!! We are going to be concerned only with the unshaded portion of the square. It also represents the fraction ½. (Which is how much of the lasagna I will want.)"

"Now let's just look at 2/3 of the unit. We do this by dividing the unit into thirds horizontally. We can shade the bottom, and the unshaded portion represents 2/3 of the unit."
He drew another square in the sand and proceeded to label it.

2/3
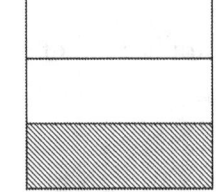

"Now," he continued, "if we put both of these pictures together, we are looking at an unshaded portion of the unit, which is 2/3 of the original ½ of the unit. We are also looking at two equal pieces out of six in the original unit. This is the same as the statement

$$\frac{2}{3} \ of \ \frac{1}{2} \ is \ \frac{2}{6}$$

or...

$$\frac{2}{3} \times \frac{1}{2} = \frac{2}{6}$$

He was drawing the picture as he spoke. He then stopped speaking, stopped writing, and simply waited for my response.

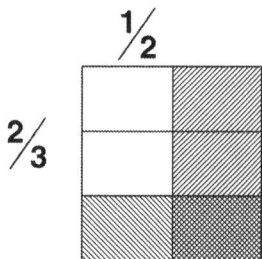

"I like that ... I really do. I can actually *see* the answer! Does this always work with fraction multiplication?"

"I sure hope so." He replied. "Let me show you a nice, and real, example, one that should satisfy both your curiosity *and* your stomach!"

Fractions for Breakfast

He stood up, stretched briefly, and proceeded into the depths of the cave. He returned a short time later carrying a handful of granola fruit bars.

"It's time for breakfast." He announced. He unwrapped one of the bars and began eating. I was getting hungry, so I snatched one of the bars from the ground, unwrapped it, and was ready to eat.

"Not so fast," he growled. "Look at that nutritious bar full of dates, nuts, and raisins. Tell me what you see."

I was startled ... and ... just wanted to eat. But ... I took a look at the bar. It was divided into six pieces, just like the picture on the ground.

"It looks just like the drawing on the cave floor. It has six pieces!"

"Correct. Why don't you eat half now and save the other half for later?"

I knew something funny was happening here, but also knew that I would play the game. It was his way of making me learn. "Ok," I said, snapping the bar in half.

We sat in silence for a minute as we both enjoyed breakfast. I wiped my hand on my shorts and looked at the other half of bar. I *did* want to eat it also.

"What do you have left?" He asked.

"You know that I have three pieces, half of the bar."

"And if you eat two-thirds of what you have left...."

"Yes!" I saw it. "If I eat two-thirds of this half I will eat two more pieces. These two pieces would be two-sixths of the original granola bar. Two-thirds of this half is the same as two-sixths of a whole bar!" All of this was spinning through my head and out of my mouth. I just multiplied fractions using a granola bar!

He smiled and listened. Finally he said, "Finish your breakfast fraction and we will look at some other examples. Would you like another mug of drink?"

Fraction Multiplication with Areas

We ate the breakfast bars, cleaned up the wrappers, and again sat on our stools. The Hermit smoothed a large area of sand on the floor of the cave. He looked at me and proceeded to explain what we were going to do.

"Let me show you another example. Here we will look at the multiplication

$$3\frac{1}{2} \times \frac{1}{3}.$$

"In order to do this I will begin with a picture that is $3\frac{1}{2}$ units long. I will divide each unit into halves so that all of the pieces in the picture are of equal size.

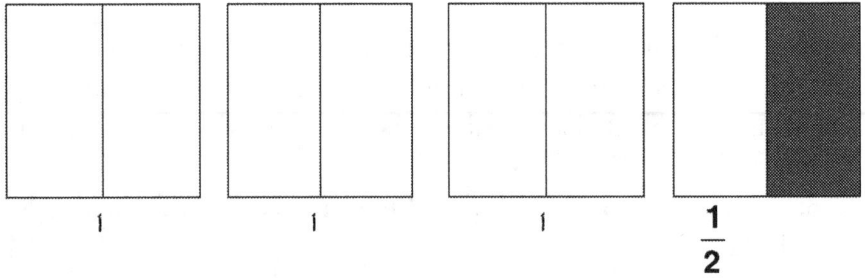

"I will now divide this picture evenly into thirds, shading the bottom two rows.

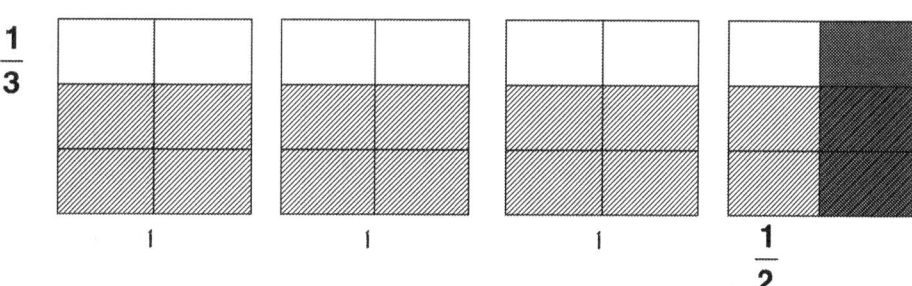

"Now I hope that you can see that the unshaded portion of our picture, which has a length of $3\frac{1}{2}$ units and a height of $\frac{1}{3}$ unit, Consists of 7 pieces, each of which is $\frac{1}{6}$ of a unit in size. Hence,

$$3\frac{1}{2} \times \frac{1}{3} = \frac{7}{6}$$

"Well...."

"I like it!" I exclaimed. "It seems so obvious now. It makes so much sense. I need to do one example for myself—I need to make sure that I really can *not only see it when you do it, but can do it for myself.*"

"I can't think of a better idea. Let's see how far you can get with this example. Hopefully ... you can get as far as the final answer!"

He scratched his head and looked up before writing. He then cleared the floor and wrote the following:

$$2\frac{1}{2} \times \frac{3}{4}$$

"There's your problem," he said. "Let's see what you can do with it."

I worked in silence. My muttering was all in my head as I began the picture that would represent the multiplication.

"I need to draw three units, and shade the last half, which I don't want to really look at.

"I need to further divide the units into fourths. I can do that and shade the bottom fourth, so I will be looking only at three-fourths.

I sat back and observed my handiwork.

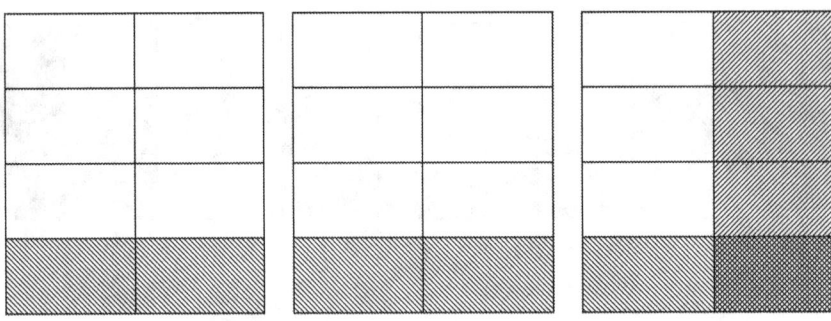

Now I was faced with the difficult part—what was the answer? I examined my wonderful drawing. I saw fifteen unshaded pieces. I also noted that each of these fifteen pieces was one-eighth of a unit. I had it!

"Here," I announced proudly, "is the solution."

$$2\frac{1}{2} \times \frac{3}{4} = \frac{15}{8}$$

"Excellent job." He said smiling. "Would you like to try just one more—should we 'kick it up a notch?'

I was not eager, but I knew I had to satisfy myself as well as the Hermit that I knew what I was doing. "Sure, lay it on me."

We cleared the floor and he wrote the problem.

$$2\frac{1}{4} \times 1\frac{2}{3}$$

"It's all yours." He said, leaning back against the wall and relaxing.

I stared at it for a time, and then began to draw and mumble under my breath as I worked.

"If I draw 3 units horizontally, and shade the right-most ¾, we will be looking at 2 ¼. I can draw 2 units vertically, and shade the bottom 1/3, leaving us looking at 1 2/3."

Here was the result of my labor:

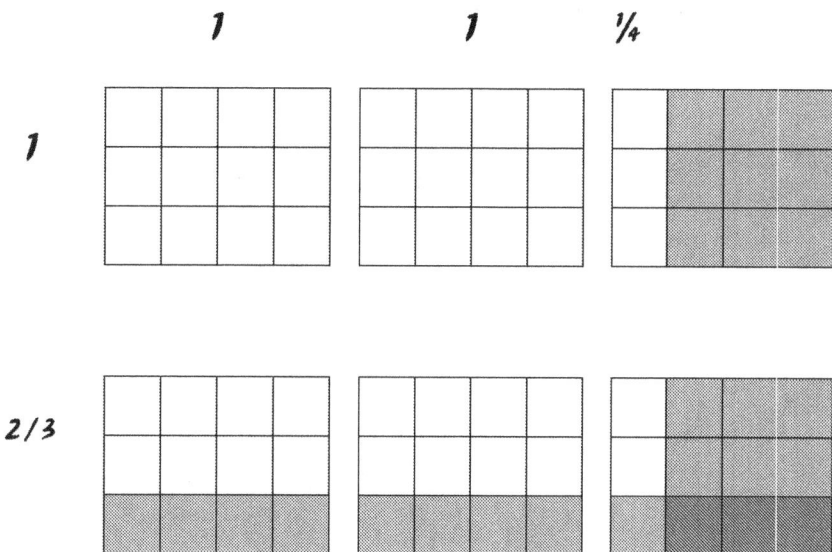

All I had to do now was to see what I was seeing … to see what I had drawn. The Hermit was being no help. He just sat silently and watched.

I counted forty-five unshaded pieces in my drawing. Each of the original units that I had drawn now had 12 equal parts. I decided to plunge in and write the answer.

$$2\frac{1}{4} \times 1\frac{2}{3} = \frac{45}{12}$$

"Excellent, excellent." He exclaimed. "See how simple things get when you *really know* what multiplication is!"

Multiplying Fractions Without Pictures

I was happy with my work, and happy to be able to find the answer. Still, I did not want to draw these pictures every time.

"Ok, there has to be a way to find the answer without drawing a picture," I said hopefully.

Yes, of course there is. He began, "We have seen that anyone should be able to multiply fractions by using a pictorial representation, and … they will also understand what fraction multiplication really is! But, you are correct that we do not always want to draw pictures—it takes too long. Now that we can 'see' what fraction multiplication represents, we can look for a method to multiply fractions without the pictures. The simple way to multiply fractions is simply to multiply the numerators and multiply the denominators. Hence, "he began writing even as he was still speaking.

$$\frac{2}{3} \times \frac{1}{2} = \frac{2 \times 1}{3 \times 2} = \frac{2}{6}$$

$$3\frac{1}{2} \times \frac{1}{3} = \frac{7}{2} \times \frac{1}{3} = \frac{7 \times 1}{2 \times 3} = \frac{7}{6}$$

$$2\frac{1}{4} \times 1\frac{2}{3} = \frac{9}{4} \times \frac{5}{3} = \frac{9 \times 5}{4 \times 3} = \frac{45}{12}$$

I was wondering how he could possibly remember all of the problems that we had tackled when he continued.

"Now we should answer the question, 'Why does this work?' Let's refer to our pictures and look at the solutions to problems. Our first problem was the following:"

$$\frac{2}{3} \times \frac{1}{2} = \frac{2}{6}$$

"and the picture looked like this:"

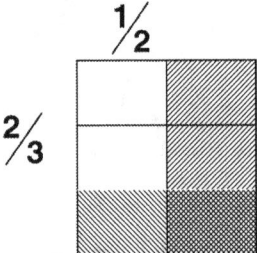

"In order to 'see' the multiplication we divided the unit first into thirds and then into halves. Hence, we then had 3 x 2, or six, equal pieces. The area we were looking at as our answer was only an area of 2 x 1, or two, equal pieces."

"Yes, I see that," I said in response. I knew he wanted one.

He then continued, "The last problem we looked at was the following:

$$2\frac{1}{4} \times 1\frac{2}{3} = \frac{45}{12}$$

and the picture of this problem looked something like this:"

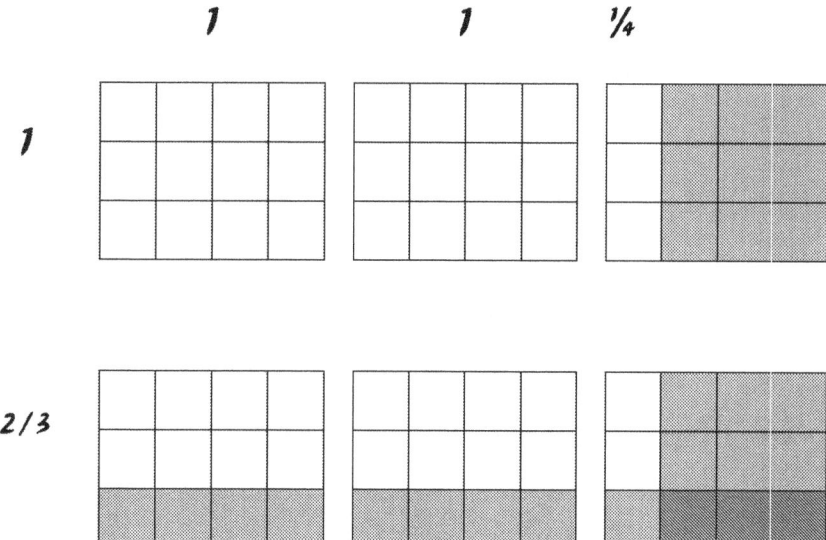

Let's look at this more closely. He began, "Notice that you began by making a length of 2 ¼ units, each divided equally into fourths. You then made a height of 1 2/3 units. Each of the *units* was now divided into 12 equal pieces (or, each piece has a size of 1/12). The unshaded area has a length of 9 and a height of 5, making it a total area of 45 (9 x 5). Thus, your answer was the following:"

$$2\frac{1}{4} \times 1\frac{2}{3} = \frac{9}{4} \times \frac{5}{3} = \frac{9 \times 5}{4 \times 3} = \frac{45}{12}$$

He continued with a question, "Do you remember our original definition of a fraction?"

I thought, then wrote and answered, "Yes it is the number of pieces you have over the number of pieces in a unit."

$$\frac{\textbf{number you have}}{\textbf{number in a unit}}$$

"Exactly," he exclaimed. "***And, the number you have will always be the area equal to the product of the numerators. The number in a unit will always be the area found by multiplying the denominators.***"

"I love this stuff!"

Does 'of' Always Mean Multiplication?

"So," he continued, "Can I assume that we are finished with this discussion ... that you are satisfied with your knowledge of multiplication?"

"Yes ... but ... er ... no!", I blurted out. "There is something that is bothering me, it is something you said earlier."

"Well," he said, "What is it?"

I paused and tried to remember the exact phrasing. This was something that had been bothering me all morning, and his statement had stuck in my head.

"Here it is," I said as I wrote.

$$\frac{2}{3} \text{ of } \frac{1}{2} \text{ is } \frac{2}{6}$$

Or ...

$$\frac{2}{3} \times \frac{1}{2} = \frac{2}{6},$$

My question is, "Does *of* always mean multiplication?"

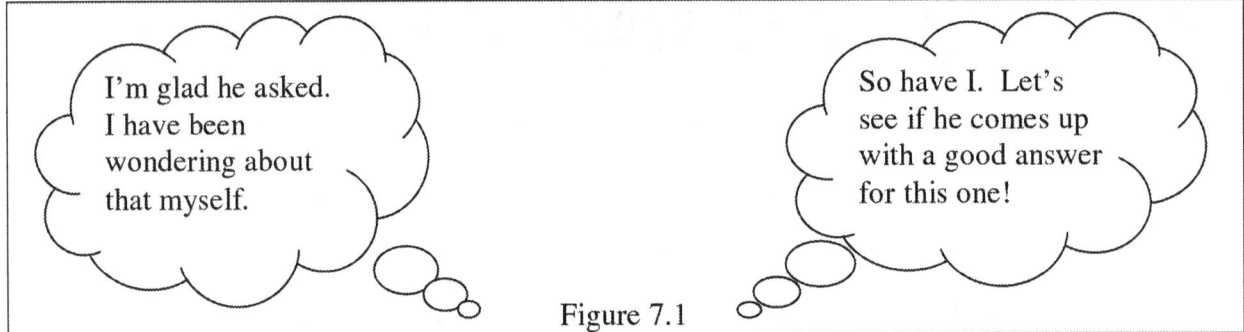

Figure 7.1

He thought about this and then replied to my question, "Many times you hear students (and teachers) of mathematics state, "*Of* means multiply." In your statement above we *did* translate *of* to mean perform the operation of multiplication. You want to know if it **always** means multiply? In order to answer let's look at some examples and determine if this statement is true."

I recall the examples he used so vividly that I remember them to this day. These are the examples that he gave that day.

Example: Five-sixths of three dozen is thirty.

Explanation I: This is similar to our previous detailed examples above. If we look at the geometric explanation of the problem, which involves an area that is thirty-six units in length and five-sixths of a unit in height, then the process of determining the answer is, indeed, a multiplication problem.

Explanation II: We can use an alternate approach to this problem. Think of our three dozen as eggs in cartons. We want to find five-sixths, or ten-twelfths, of the eggs in each carton. Thus, we find ten eggs in each carton, or 30 eggs in the three cartons. Once again this problem can be solved by multiplication.

$$\frac{5}{6} \times \frac{36}{1} = \frac{5 \times 6 \times \cancel{6}}{\cancel{6}} = 30$$

Example: During a game of horseshoes, two of four horseshoes landed in the *pit* (the fond name of the landing area in the game of horseshoes). How many horseshoes do you expect to find in the *pit*?

Explanation: It would appear that sometime during a game, four horseshoes were thrown and two of these landed in the *pit*. There does not appear to be any mathematical operation stated, and certainly the statement 2 x 4 = 8 has no relevance in this example.

Example: In the course of a softball game a pitcher struck out 15 of 21 batters she faced. How many did *not* strike out?

Explanation: If she struck out 15 batters, those that did not strike out is found by subtraction: 21 − 15 = 6. Notice that if we multiply, 15 x 21 = 315! This certainly becomes a ridiculous answer if you know anything about the game of softball.

He then finished the conversation that day with the following statement. This one has always stayed with me.

"Obviously the word *of* can carry a variety of mathematical meanings (or mean nothing mathematical at all, as in our horseshoe example). **You must determine, using your knowledge of mathematics and your knowledge of language, what the word *of* really means**."

Session 8—Fraction Division

The Meaning of Fractional Division

There are times, especially when it is a steaming hot summer day, when it is simply enough to sit at the entrance to the Hermit Cave and look across the valley and watch the heat rise from its floor. This happened to be one of those days. I was simply sitting at the cave entrance watching ... nothing. In the shadow of the cave entrance I was probably twenty-five degrees cooler than if I moved two feet into the sunlight. Thoughts and ideas were slipping into and out of my like fish darting into and out of sight on a coral reef. The Hermit was not here when I arrived so I simply sat and waited. Eventually I was able to make out his shape scuttling across the valley floor, carrying a huge sack on his back. He looked like a ... Hermit Santa Claus! He finally arrived at the cave and we exchanged greetings. It was as if he knew I would be here and there was absolutely no surprise in his voice upon seeing me.

"I need a cold drink," he said after our hellos. "Can I get you one also?"

"That would be nice."

He disappeared into the depths of the cave and returned with two large tumblers of cold water. I knew that he kept the water in clay pots in the rear of the cave, and it always seemed to be the most refreshing I had ever tasted.

"What have you been thinking about?" He could be very blunt with his questions.

"Oh, a lot of things. One of them was division of fractions. Since we discussed fraction multiplication and I now know what it means ... and ... it even makes sense, I began thinking about fraction division. I know I should 'invert and multiply' to get the answer, but that's not enough for me now. I can correctly simplify problems but want to know what I am doing when I am doing it. Watch."

I picked up a stick, smoothed the sand at the entrance, and wrote the following:

$$3 \div \frac{2}{3}$$

$$= \frac{3}{1} \times \frac{3}{2}$$

$$= \frac{9}{2}$$

$$\frac{7}{3} \div \frac{1}{2}$$

$$= \frac{7}{3} \times \frac{2}{1}$$

$$= \frac{14}{3}$$

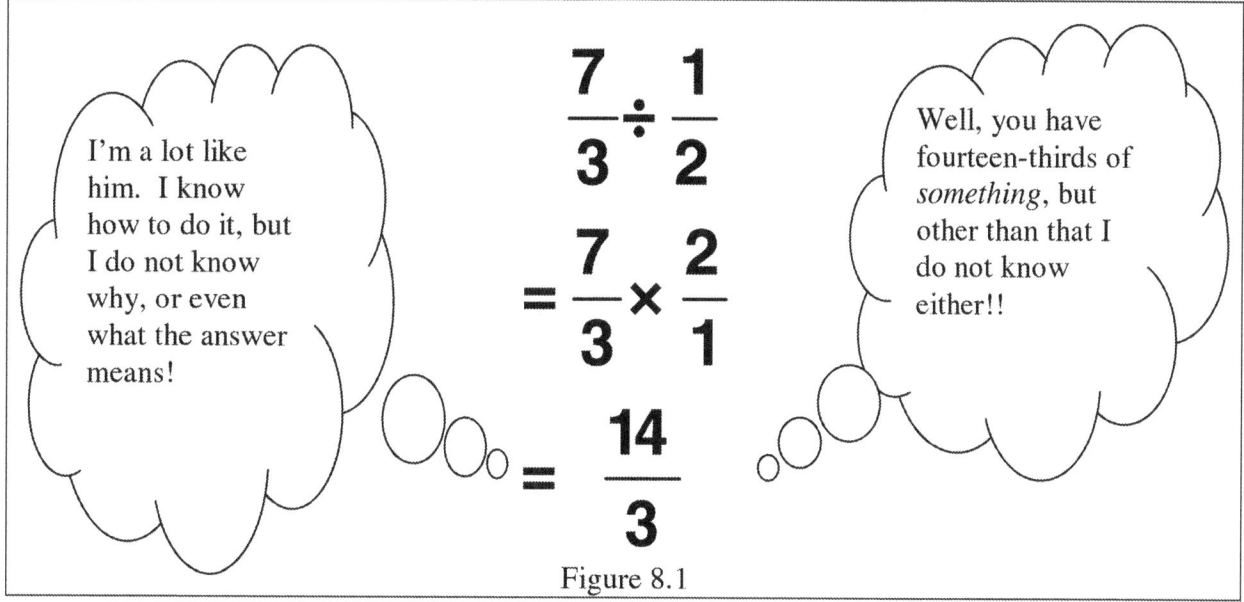

Figure 8.1

"See," I began, "I can do the problems correctly. But … I don't know … I'm not exactly sure … what it means! Does anyone know what these mean?"

He thought for a few seconds before answering. Finally, after staring at my two examples and taking a long drink of the nectar in his tumbler, he spoke.

"Let's look at your first example, three divided by two-thirds. Let's suppose that I have three boxes of *things* in the back of the cave, and for storage purposes I need to *divide* the three boxes into smaller pieces. Further, we will divide the three boxes up into pieces each two-thirds of a box in size. The question becomes 'How many two-thirds are there in three?' This is the question we are asking here."

"So, how do we do it?"

"Well, I really don't have the boxes … but we can draw some. I'll do that here."

I watched as he cleared a space and drew his picture.

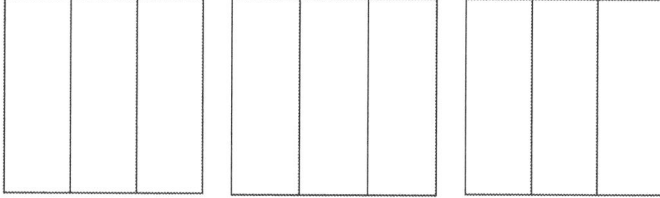

"Notice that I've taken the liberty of showing you the 'thirds' in each of the boxes. *We started with three, and now we have nine pieces. Each of these nine pieces has a size of one-third.* Now I need to divide them into groups of two-thirds … or … find out how many groups of two-thirds of a box are in three boxes. Watch:"

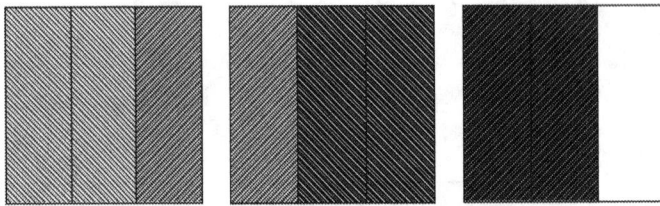

"I now have four groups of two (which is four groups of two-thirds) and one left over. Since each group has two parts or two pieces, my leftover is a half of a group. My answer matches your answer. Watch."

$$\frac{9}{2} = 4\frac{1}{2}$$

"Now we can look at you second example. You wrote the following:"

$$\frac{7}{3} \div \frac{1}{2}$$

$$= \frac{7}{3} \times \frac{2}{1}$$

$$= \frac{14}{3}$$

"OK, let me draw another picture. This will begin with seven-thirds units:"

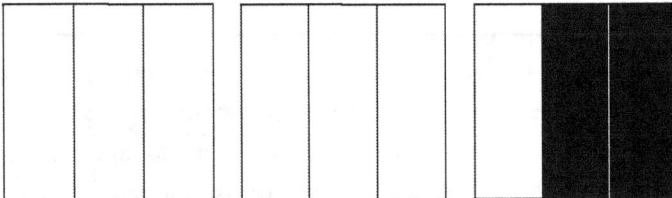

"Notice that I have taken the liberty of dividing the units into thirds. I will only consider 7 of these thirds. I have seven pieces to work with. Now … you want to divide these into halves. I will divide all of these into half of what their size presently is. It will look like this:"

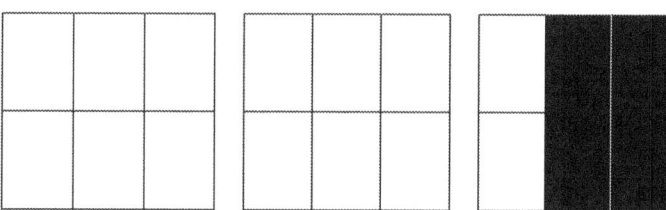

"Notice that when I divided my seven pieces into halves *I produced two times as many pieces as I originally had*. Now, your question stated, 'How many halves are in seven thirds.' I need to find the number of halves in my picture. I will shade the halves and see how many of these I have."

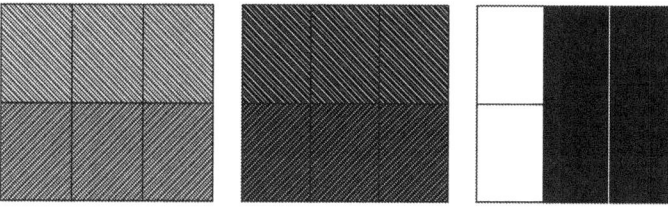

"Notice that each '*half*' has three pieces. That means that the two unshaded pieces represent two-thirds of a half, or two-thirds of my *new* unit. My answer is the following:"

$$\frac{7}{3} \div \frac{1}{2} = 4\frac{2}{3}$$

"This is exactly the same as your answer. Look:"

$$\frac{14}{3} = 4\frac{2}{3}$$

So ... What *is* Division of Fractions?

I was beginning to see what he was saying, but was still bothered by *what*, exactly, we were doing. I had to ask, "I think I see what you did, but ... *what is division* when dealing with fractions?"

"Division is division," he replied. "It doesn't matter if you are dealing with whole numbers or fractions or decimals or any combination of these. Division means you want to *find equal parts* of whatever you have. The first example you gave me stated this." He proceeded to point to my original question.

$$3 \div \frac{2}{3}$$

"Recall that we wanted to divide three into equal parts, each of which was two-thirds of a unit in size. We found the following:"

$$3 \div \frac{2}{3} = 4\frac{1}{2}$$

"Your second example was the following," he said, pointing to what I had written on the floor.

$$\frac{7}{3} \div \frac{1}{2}$$

"What you were really asking was, 'How many halves are in seven-thirds?' Our answer was the following:"

$$\frac{7}{3} \div \frac{1}{2} = 4\frac{2}{3}$$

"We found that there were four and two-thirds *halves* in seven-thirds!"

I was still a little fuzzy on the concept of division. But ... I had to admit ... what he said *did* make sense. I needed to see for myself if what I was thinking was correct. I asked, "Can I do an example for myself, just to see if what *I think I see* here is correct?"

"Definitely," he replied. "You can do as many as you wish. Simply make sure that when you do a problem you know what you are doing and why you are doing it."

"Good. Here is the problem—let me solve it myself."

"As you wish...."

I quickly cleared the area before me and wrote the following:

$$\frac{3}{4} \div \frac{2}{3}$$

"OK," I said, thinking aloud. "I need to first have three-fourths of a unit, then divide that into thirds." I drew the following:

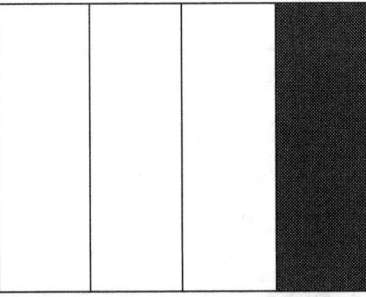

"Now," I continued, "I simply have to divide that into thirds."

"OK," I said. "I have no idea what to do next ... or even if what I have drawn is *what I wanted to draw!*", I said, throwing my arms up in disgust! My stick flew out of my hand and out the entrance of the cave. It disappeared in the harsh sunlight.

"You have done a good job," Hermit said, encouraging me. "Now is the time you need to *think*. How many pieces are now in your unit?"

"I have twelve pieces drawn."

"And ... one-third of that is how many?"

"That would be four pieces."

"So ... two-thirds would be ..."

"Eight!"

"Correct. Shade in eight of the pieces. That will represent two-thirds."

I quickly found another stick and shaded the eight pieces of my unit.

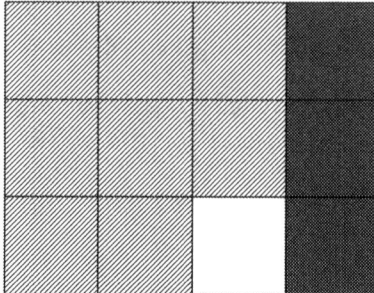

"So," he said, "what do you see now?"

Hesitatingly, I said, "I see ... one ... and ... one left over. Since my unit now has *eight pieces*, the left over piece must represent one-eighth. The answer is one and one eighth."

The Hermit didn't say anything. He simply scribbled the following in the sand:

$$\frac{3}{4} \div \frac{2}{3} = 1\frac{1}{8}$$

"Your division problem shows that you would have one and one-eighth pieces which are two-thirds in size from a unit which is three-quarters in size."

"I see this.... I can see all of these examples. I still need to know why I *invert and multiply!*"

Invert and Multiply

"I mean," I continued, "I can almost see where the numbers are coming from. I can see the relation between your pictures and the answers to the problems. But ... I'm still not sure *exactly* what is happening here."

There were a few seconds of silence as we both stared at the floor and all of the figures and drawings that were there. I knew that we were looking at these things differently. I was sitting, trying to see all of the various relationships that were present. I knew that the Hermit already saw these relationships, and was trying to develop a strategy that would allow me to also see them. Finally he spoke.

"You recall, when we first began examining fractions, one of our first explanations of what, exactly, a fraction is, I hope. We said that a fractional form of a number could be defined as the following." He proceeded to smooth the sand and draw on the floor.

$$\frac{\textbf{the number of equal pieces you have}}{\textbf{the number of equal pieces in a unit}}$$

"Your first example was the following:"

$$3 \div \frac{2}{3}$$

"We started with three equal pieces, each of which was one unit in size. That is why you wrote, on the next line, $3 = \frac{3}{1}$. When you wanted to divide this into equal parts we needed thirds. We ended up with 3 x 3 = 9 equal parts. Recall the picture." He proceeded to draw yet another picture as he added to the problem statement.

$$3 \div \frac{2}{3}$$
$$= \frac{3}{1} \times \frac{3}{2}$$

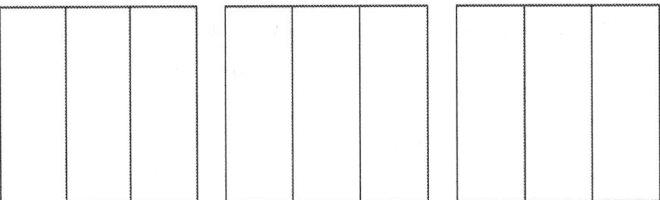

"Now, we group these thirds *two at a time*. Recalling our definition of a fraction we arrive at the following conclusion:"

$$3 \div \frac{2}{3}$$
$$= \frac{3}{1} \times \frac{3}{2}$$

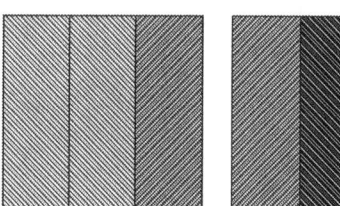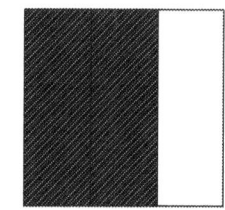

$$\frac{\text{the number of equal pieces you have}}{\text{the number of equal pieces in a unit}} = \frac{9}{2}$$

I was stunned. It was all so obvious now. There were the *nine pieces* and I knew that there were *two in each new unit*. He interrupted my joy with a further explanation.

"When we divide we always create new units. We are always re-grouping when we divide, whether it is whole number division or fraction division. Do you really see what we just did here?"

"Yeah, I think I do. But … can we do another example, just to make sure?"

"Definitely. Let's do the second one that you earlier proposed." He cleared the sand and quickly scribbled the problem and picture. I had no idea how he remembered this example, which I had already forgotten.

$$\frac{7}{3} \div \frac{1}{2}$$

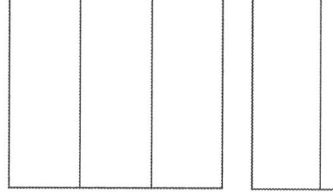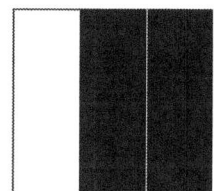

"Now," he proceeded, "we are looking at seven equal pieces, each of which has a size of one-third."

"I agree."

"And … we want to know how many halves are here."

"I agree again." I said, nodding my head.

"Then let's look at halves. Here's the picture."

$$\frac{7}{3} \div \frac{1}{2}$$

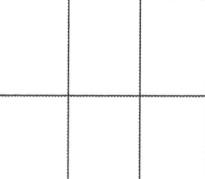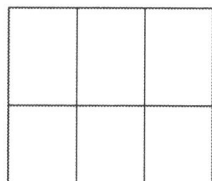

"Now," he began after a few seconds, "How many equal pieces are there that we are looking at?"

I counted and then replied, "There are fourteen."

"And ... how many of these make up are *new unit* of one-half?"

"That would be three."

"Good, we're finished here."

$$\frac{7}{3} \div \frac{1}{2}$$

$$= \frac{7}{3} \times \frac{2}{1}$$

$$= \frac{14}{3}$$

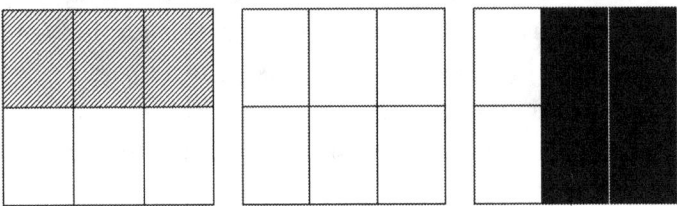

I couldn't believe it. I actually knew why I did division the way that I did it. All of the magic was gone. Well, *almost* all of the magic was gone. There was still the magic inside my head that said that I understood what I was doing. I needed more. I had to do one problem, start to finish, by myself.

"One more, just for me," I begged.

"I knew you would ask. I just happened to be thinking about a nice example. You can do it first; then, I will try. Here is your problem."

Once again the sand was smoothed, a stick retrieved, and symbols appeared on the cave floor.

$$2\frac{1}{2} \div \frac{3}{4}$$

I stared for a short time, muttering to myself as I did so. "First, I need to see two and a half."

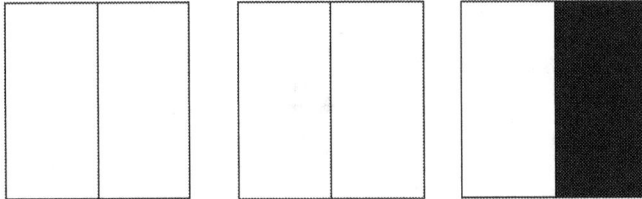

"Now I need to divide that into fourths."

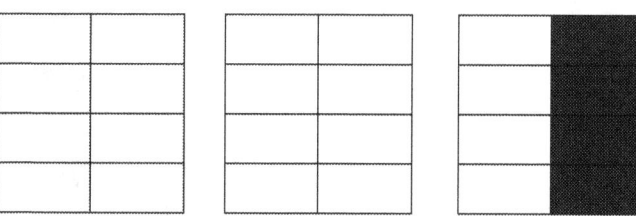

Still muttering to myself, "OK, I have twenty equal pieces. Now I need to see how many of these make up three-fourths. I can shade *one* three-fourths and find out."

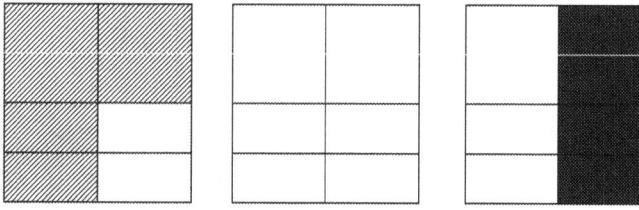

"There are six equal pieces in my three-fourths unit. There are twenty pieces total. Here I go, for all of the marbles." I quickly wrote what I hoped was the correct answer.

$$2\frac{1}{2} \div \frac{3}{4} = \frac{20}{6}$$

"There, the answer is twenty over six," I said proudly.

"Well," Hermit replied, "We can check it. Watch."

$$2\frac{1}{2} \div \frac{3}{4}$$
$$= \frac{5}{2} \times \frac{4}{3}$$
$$= \frac{20}{6}$$

"Yessssss!" It hissed from my lips as I let my pride show.

"You recall my saying that I also wanted to do this problem," Hermit stated.

"But we've already done it! Why do it again?"

"Humor me, let me have my fun."

"If you insist." I reached for my still cool nectar, having become parched by all of the talking and deep thinking. A good, long drink would do me good.

"I'll begin with the same picture as you did," he stated.

$$2\frac{1}{2} \div \frac{3}{4}$$

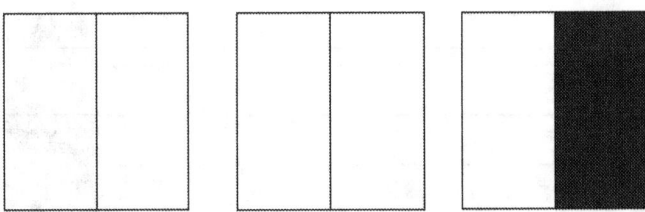

"I also need fourths, so I am going to divide my units into fourths."

$$2\frac{1}{2} \div \frac{3}{4}$$

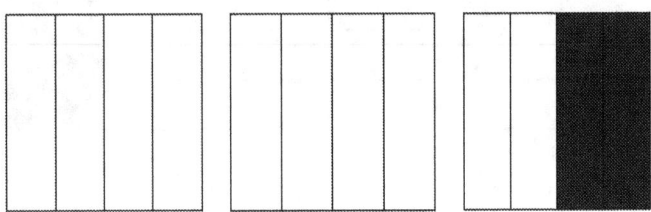

"I, like you, will shade a *new* unit of three-fourths."

$$2\frac{1}{2} \div \frac{3}{4}$$

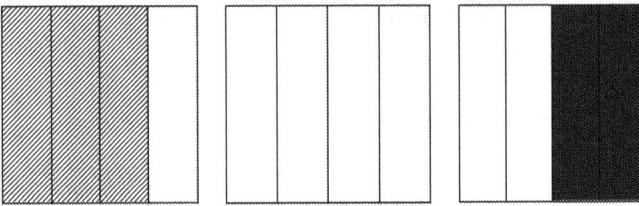

"My new unit has three equal pieces, and I have a total of ten pieces. Here is *my* answer."

$$2\frac{1}{2} \div \frac{3}{4}$$
$$= \frac{10}{3}$$

"How did you … what did you … wait! That's not the same as the answer I got," I said, exasperated.
"Ah, but it is. Let's compare yours to mine."

$$\frac{20}{6}$$

$$= \frac{10 \times \cancel{2}}{3 \times \cancel{2}}$$

$$= \frac{10}{3}$$

"Wow," was all I could say.

Session 9—Fraction Addition

Introduction

I had been sitting at the entrance to the Hermit Cave for quite some time before the Hermit came out from the depths.

"It's good to see you here my young friend. I was afraid that I had scared you with our previous discussion of fractions." He stated bluntly.

"No," I said in reply. "Quite the contrary—I have come up with some questions that I would really like answered."

"OK. I will assume that these questions have to do with fractions. I will try my best to answer them for you."

"Well, I'm not sure where to start, but here goes," I began.

"We have been discussing fractions and the operations of arithmetic, and we have done so with other numbers. I guess my main question is about *common denominators*. I don't seem to know why we *need* these to add fractions, and when we need to use them. This confuses me and it makes me less successful than I might otherwise be. Addition is, I think, finding the sum or total of two numbers." I stated.

"Well," he said, "I guess if we are going to discuss addition of fractions then common denominators is a good of a place as any to begin. But ... before that discussion we should review what we know about addition. Do you remember what our definition of addition was?"

"Yes, I do. We said that we had to have some things, and to these we would put more on the pile. The act of placing more on the pile is what we call addition."

"I can go along with that. Let me give you a problem and see if you an come up with the correct answer. Let me write it in the sand."

He proceeded to pick up a stick and write the following in the dust.

$$6 + 5$$

"There is your addition problem. Let's see what you can do with it.

This was all becoming so familiar. It was a simple problem and I knew that when answering it I would be walking into a trap. But ... I also knew that by walking into the trap I would be beginning my process of learning. I picked up a different stick and wrote the following:

$$6 + 5 = 11$$

"There, the answer to six plus five is eleven." I stated with confidence.

"That would be the usual answer people would quickly give," said the Hermit.

76

"However, let me explain to you where, exactly, I came up with this problem. I saw something in the store this morning and was not sure whether I had enough money to buy it. I searched my pockets and found 6 quarters and 5 dimes. Now … does your answer of eleven tell me how much money I had?"

"Well … " I stammered. "We do know that you had eleven coins, but, no, we do not know how much money you had."

"One of the goals of addition, or any other mathematical operation, is to *simplify* a problem, give us more information, and make the situation more understandable in an easier form. What we have here is the knowledge that we have eleven coins, but if I had not told you, we would not know their denominations, how many of each, or the total amount of money that I had."

"So, … how do I add these two amounts?"

"Well, most of the time we make an assumption that the two quantities are of the same size, or in the case of money, of the same denomination. Here we have two different size coins, quarters and dimes. The best that we can do is to say that we have six quarters and five dimes—adding them gives us no additional information.

Can you add Two plus Two?

"Here is another example," he continued. "The classic example for being the simplest math problem that there is: What is the answer to two plus two?"

"Well," I started, "I want to say four, but…."

"But now you know what you don't know, and that is if the items are all the same!"

"That is correct. ***I know what I do not know***!!"

"Absolutely. You have two of something (maybe bags of candies) and two of something else (maybe pecans), but you do not know how many of anything you have."

"So, … " I interjected, "I do not know if I have four of something or have two of one thing and two of another!!"

"So, … " the Hermit responded, "what is the correct answer to the question of two plus two?"

"Well, ***the most incorrect answer*** would be four, since I don't know anything about the items. Probably … the most correct answer is that I don't know the answer to two plus two, unless you give me more information."

"That is precisely the most correct answer you could give. If you would have told me that the answer was four (which is what most people would reply with) I would have to tell you that you gave me the absolutely most incorrect answer."

Common Denominations

"Now," he continued, "let us return to our original problem, one which we have not yet solved. Recall that I had in my pocket six quarters and five dimes, and wanted to know if I had enough money to purchase something for $1.85. We could not add six quarters to five dimes, but … maybe … if they had something in common, *a common denomination*, we could find a way to add them. Can you think of any denomination of coin that dimes and quarters have in common?"

"Well, they both contain pennies, or cents."

"That is correct. Each dime contains 10 cents and each quarter contains 25 cents. Let's change both of our coin types into pennies or cents, and maybe we will be able to see how much we have."

The Hermit quickly began to smooth the sand with a small broom that always seemed handy when he needed it, and then he began to write in the sand.

$$6 \text{ quarters} = 6 \text{ (25 cents)} = 150 \text{ cents}$$
$$5 \text{ dimes} = 5 \text{ (10 cents)} = 50 \text{ cents}$$

"So there is what we have in cents," he observed. Let us now see if we can find the *total amount of money* that was in my pocket."

$$6 \text{ quarters} + 5 \text{ dimes} = 150 \text{ cents} + 50 \text{ cents}$$
$$= 200 \text{ cents (or \$2.00)}$$

"I had 200 cents, or \$2.00, in my pocket. I could have purchased the item which cost only \$1.85!"

"So you could have purchased the item, and we only needed to find a find a *common denomination* to determine this."

"Yes, that is true."

"OK, my curiosity has gotten the best of me. Is there another way we could have arrived at the answer?"

"Well … yes … there is another way. Can you think of **another** *common denomination* for both dimes and quarters, one different from pennies?"

"Yes, dimes and quarters are both composed of nickels—each dime has two nickels and each quarter has five nickels."

"Exactly!! Using the same method as before, we can determine the number of nickels in each quantity."

He proceeded, once again, to sweep the sand and write the following:

$$6 \text{ quarters} = 6 \text{ (5 nickels)} = 30 \text{ nickels}$$
$$5 \text{ dimes} = 5 \text{ (2 nickels)} = 10 \text{ nickels}$$
$$6 \text{ quarters} + 5 \text{ dimes} = 30 \text{ nickels} + 10 \text{ nickels}$$
$$= 40 \text{ nickels}$$

"So," he interjected, "we got an answer of 200 cents, and I got an answer of 40 nickels. Which of us is correct—or are both of us correct?"

"OK", I answered, "either one of us is wrong or 200 cents is equivalent to 40 nickels. We looked at equivalent quantities earlier, and should be able to determine if these are equivalent."

"Good, let's look at them. Each nickel contains 5 cents. So we have the following:"

$$40 \text{ (nickels)} = 40 \text{ (5 cents)} = 200 \text{ cents}$$

"Wow," I exclaimed, "That is really neat. There were two different ways to solve the same problem.

"But … we still have to address the problem of the addition of fractions … I think."

"You are almost correct," replied the Hermit. "We have and we have not addressed the problem of fraction addition. When we added quarters and dimes, we added fractions of dollars. Not only did we add fractions of dollars, but we did it in two different ways, by adding cents and adding nickels.

Common Denominators

"Let's recall our definition of a fraction," stated the Hermit as he smoothed the sand and began to write.

the number of equal pieces you have
the number of equal pieces in a unit

"What we need in order to add fractions is to have all of the pieces the same size. We have done this here today by making all of my coins equivalent to pennies. Recall that I had 6 quarters, which was the same as 150 pennies, and I had 5 dimes, which was the same as 50 pennies. We could not add the quarters and dimes, which were different, but were able to add the pennies, which had the same value."

"So," I stated, "we found a common denomination. This is the same as a common denominator?"

"Well, cents are a part of a dollar, and dimes are a part of a dollar, and quarters are a part of a dollar.… The number of cents in a dollar is 100, the number of dimes in a dollar is 10, and the number of quarters in a dollar is 4. We could write the following as fractions:"

$$1 \text{ cent} = \frac{1}{100} \text{ of a dollar}$$

$$1 \text{ dime} = \frac{1}{10} \text{ of a dollar}$$

$$1 \text{ quarter} = \frac{1}{4} \text{ of a dollar}$$

"Notice that the denominators are all different. In order to add the amounts we had we needed to make the denominators the same. Let's look at what we had. A quarter is $\frac{1}{4}$, or $\frac{25}{100}$ of a dollar. A dime is $\frac{1}{10}$, or $\frac{10}{100}$ of a dollar. We wanted to add 6 quarters (or $6 \times \frac{25}{100} = \frac{150}{100}$) to 5 dimes (or $5 \times \frac{10}{100} = \frac{50}{100}$) then we had the following:

$$6 \text{ quarters} + 5 \text{ dimes} = \frac{150}{100} + \frac{50}{100}$$
$$= \frac{200}{100}$$

"And," he said, "this is the same answer we arrived at earlier!

"This is what is known as 'finding a common denominator.' *It is necessary to make all of the pieces the same size in order to add* (or subtract) them. It is not that difficult, but you should (one more time) *know why* you do something rather than just doing it because someone tells you to."

MORE Common Denominators

"In the examples that we have examined we were looking at fractions that we were familiar with, as they involved coins," he stated. "Let's look at some other examples that we may not be as familiar, but we will find that they are just as simple. Here is a very easy example, and we will work our way up to more difficult problems."

He smoothed the sand at his feet and wrote the following:

$$\frac{2}{3} + \frac{1}{4}$$

"Obviously," he started, "the two denominators are not the same. We can think of having two-thirds of a box of chocolate and we want to add to that one-fourth of a box of the same chocolates. The question becomes one of *making all of the pieces the same size*!

"There are a variety of different methods available for use when adding fractions," said the Hermit. "I always use the same method because I find that it is simple for me.

"I am simply going to multiply two-thirds by four-fourths and multiply one-fourth by three-thirds. Since I am multiplying both fractions by *one* I am not changing the value of the fractions, simply the size of the pieces! Watch:"

$$\frac{2}{3} \times \frac{4}{4} + \frac{1}{4} \times \frac{3}{3}$$

$$= \frac{8}{12} + \frac{3}{12}$$

$$= \frac{11}{12}$$

"And this always works?" I asked.

"I don't seem to recall an instance in which it has failed me. Many people use a method of finding the Least Common Denominator, or LCD. You have to search the fractions, do numerous factorings, and then need to decide what the common denominator indeed, is. A lot of people simply end up scratching their head and doing nothing else.

"Of course, if you *do not find* the LCD you may end up having to simplify your answer before you are finished."

"Can we look at one like that so that I can see the difference?"

"Sure. I will do this problem the same way as the other one, but at the end I will have to simplify my answer in order to compare the two equivalent answers.

$$\frac{3}{4} + \frac{1}{6}$$

$$= \frac{3}{4} \times \frac{6}{6} + \frac{1}{6} \times \frac{4}{4}$$

$$= \frac{18}{24} + \frac{4}{24}$$

$$= \frac{22}{24}$$

"Now we will need to simplify the fraction. It has been a while since I've seen you do this—would you like to give it a try?"

"Why not! I could use the practice and I *should* remember how to do this. I will factor both the numerator and denominator into primes, then see what they have in common."

$$22 = 2 \times 11$$
$$24 = 2 \times 2 \times 2 \times 3$$

"Now I will have the following simplification:"

$$\frac{22}{24} = \frac{\cancel{2} \times 11}{\cancel{2} \times 2 \times 2 \times 3}$$
$$= \frac{11}{12}$$

"That is very good! You have not forgotten your simplification skills. Would you like to try a more difficult problem, perhaps do it from beginning to end?"

This was a challenge that I could not resist. "Why, of course!"

"OK, here is your problem." He smoothed the sand and wrote the following in the smoothed area:

$$\frac{1}{4} + \frac{5}{6} + \frac{2}{3}$$

"Wow!!" I stated. "If I am going to do this problem the same way as the others, I will have to make the denominators the same. The easiest way that I can see is to multiply 4 x 6 x 3. This will make them all the same. I can do this by multiplying 4 times 6 and 3, by multiplying 6 by 4 and 3, and multiplying 3 by 4 and 6. My work will be the following."

I smoothed the sand 'ala the Hermit', and wrote the following.

$$\frac{1}{4} + \frac{5}{6} + \frac{2}{3}$$
$$= \frac{1}{4} \times \frac{18}{18} + \frac{5}{6} \times \frac{12}{12} + \frac{2}{3} \times \frac{24}{24}$$
$$= \frac{18}{72} + \frac{60}{72} + \frac{48}{72}$$
$$= \frac{126}{72}$$
$$= \frac{\cancel{2} \times \cancel{3} \times \cancel{3} \times 7}{\cancel{2} \times 2 \times 2 \times \cancel{3} \times \cancel{3}}$$
$$= \frac{7}{4}$$

"Very good, my young friend. Someone has been teaching you well!!"

"Well, I have been lucky enough to have a variety of good teachers in mathematics. What I am seeking now is not knowledge of 'why' things work the way that they do, but knowledge of 'how' they work.

"For instance, I now can see the need for common denominators in the addition of fractions. I now want to know the distinction between addition and subtraction of fractions. IS there a difference (Wow, I just made a math pun!!) or are the applications the same?"

Fraction Subtraction

"Well, I hope that you are not asking me if addition and subtraction are the same! What I think you want to know is if there are any major differences between the two operations when there are fractions involved."

"Sure, that is what I wanted. I mean, I know that there are differences between addition and subtraction, as we saw that with whole numbers. I guess what I was asking was if there are any major differences between the mechanics of addition and subtraction when these operations involve fractions."

"*The mechanics of these operations*—I'm impressed with your language! There are a couple of differences between addition and subtraction of fractions, but they are minor differences. You have, however, made an important observation here today. You now know, from our prior discussions, that there are differences between addition and subtraction, and I will assume that you recall that there are also similarities between the two operations. Recall that long, long ago …"

"In a place far, far away …"

Hermit Soliloquy

"No, actually we were right here. We concluded one day that we could perform addition in order to arrive at the answer to a subtraction problem because the operations were so closely related. You also want to now know how closely related the operations of addition and subtraction are when applied to fractions. You have applied a very important facet of mathematics to your logic here. Do you know what that is?"

"Well, … I'm not sure."

"It is the fact that mathematics is a cumulative entity. What I mean by this is that it continuously builds on itself as you gain more knowledge of it. What it also means is that if there are any deficiencies in your math knowledge it will make the learning much more difficult. This makes mathematics different from other subjects that you might wish to study. In many of these you can study one aspect with only minimal knowledge of the other aspects.

"You can, for instance, become an expert on the subject of the American Civil War or of the Spanish Civil War and still know very little about the other. In terms of sciences, you can get a degree with a specialization in botany or zoology, and yet, again, know very little of the nuances of the other.

"In mathematics, however, every aspect is important when one considers obtaining more knowledge of the subject. For instance, we, a long time ago, studied the addition of whole numbers. After that we delved into the operation of subtraction in a similar way. In fact, we found that addition and subtraction were closely related—closely related enough that we could perform subtractions by relating them to addition problems and adding rather than subtracting. We were able to, if we wished, to eliminate the *process* of subtraction (as most people perform it) completely, including borrowing!

"We also found, when examining the operations of multiplication and division, that they had a close relationship to the operations of addition and subtraction, respectively. It was only through all of these building blocks that we were able to come up with a reasonable and efficient meaning for what is fondly known as *The order of operations.*

"It is unfortunate that many people consider the things we have looked at as being 'trivial' and simply 'number crunching' exercises. They look at the order of operations or at fraction division as *magical mathematics*. I have seen many times on the internet, when searching for the 'order of operations', that PEMDAS was invented by men a long time ago so everyone would get the same answer! It makes you wonder what people who do not speak English do when faced with the same problem! When searching for fraction division, I am forever being told to 'invert and multiply' without being told why, or what the answer means!

"Unfortunately, many people do not look into, in any depth, the concepts that we have been discussing. They simply assume that because these are the 'basic' concepts of mathematics, they are not as important and need not be stressed as much as what they consider to be more complex or advanced mathematical ideas. Even a genius like Sir Isaac Newton (who discovered the laws of universal gravitation and the Calculus) noted that his work was trivial compared to the scientists and mathematicians that preceded him. He noted that without the knowledge of their 'basic' discoveries he would never have been able to discover the things that he did.

"These basics of mathematics that we have been discussing are much like the foundation of a building. Without a strong foundation it matters little what type of structure you attempt to build upon it. It will be weak in storms, will not last for very long, and will give you innumerable headaches. It is very similar in mathematics: You will only succeed as much as your knowledge of the basics will allow you. When your knowledge of mathematics is based on a firm and complete understanding of the basic concepts, this same knowledge will be able to grow more rapidly and more completely.

Fraction Subtraction—Continued

"I'm sorry, I got carried away with my thoughts. Where, exactly, were we in our discussion?"

"We were beginning a discussion of fraction subtraction," I observed. "We were going to determine what the similarities and differences were between subtraction and our previous topics."

"Ah, now I remember. Similar to our work with whole numbers, when we perform an addition we will put more items in our pile, and when we perform a subtraction we will remove some of the items from our pile. With fractions, however, we have to be sure that the all of the pieces we are working with have the same size, or have a …"

"… common denominator!"

"Exactly, my friend. As a simple example, and since I baked some bread yesterday, the Tandoor is still hot, and I was thinking that I might make a pizza tonight."

"The Tandoor is that big rock-like thing in the rear of the cave?"

"Well, actually it is clay and fired by charcoals. It is wonderful for all kinds of baking. Let's suppose that I do make the pizza, and when we are done eating there are still five of the original twelve pieces of pizza left. Now, if I arise before you in the morning and eat three of the pieces, how much will there be left for you?"

"O.K. The unit, or pizza originally had twelve pieces. There were five left in the morning and you ate three. I would get this:

$$\frac{5}{12} - \frac{3}{12} = \frac{2}{12}$$

"I would have two pieces to eat!"

"Very nice. There is one thing I would like to point out to you (again). All of our fractions were left with denominators of twelve. If I reduce them, the amounts or values will remain the same, but we would see the following:

$$\frac{5}{12} - \frac{1}{4} = \frac{1}{6}$$

"This is an example of losing information when you reduce or simplify fractions. The bottom equation does not tell me how many pieces of pizza I ate in the morning. The bottom equation does not tell me how many pieces of pizza *you* had to eat in the morning. It is also much easier to compare fractions when they have a common denominator. Comparing three-twelfths to two-twelfths is much easier than comparing one-fourth to one-sixth."

"I know, and you cannot tell me whether it is better to simplify the fractions or not because it is dependent upon the circumstances!"

"Do you realize that you are really beginning to *teach yourself.* There may come a time soon when you will no longer need me. You will be able to formulate your own questions and obtain reasonable answers. That is when you *know* that you are really learning mathematics. Would you like to try one more subtraction problem now, just to see how proficient you are?"

"Why, sure, I would like to try one more."

"Good. Make up your own problem, solve it, and check the answer. You are on your own now. I have to make the pizza crust … and I have been working on a problem here and it needs some private thinking. A group of mathematicians has just proven the existence of a new mathematical entity. To the layperson it looks much like a huge spider web, and with no strings attached. I thought I had a wonderful and elegant proof here, but it seems that the margin of the paper in which I was working is not large enough. I will have to retreat to the innards of the cave and find more paper. Begin, do not wait for me to return.

978-0-595-44417-5
0-595-44417-2